BIRDS IN THE YORKSHIRE MUSEUM

The illustration on the front cover shows the Saker *Falco cherrug* mount that was donated to the Museum by the Al-Areen Wildlife Park of Bahrain.

North Yorkshire County Council

BIRDS IN THE YORKSHIRE MUSEUM

Michael Denton, MBOU, FRES

ISBN 0 905807 10 3

Published in February 1995 by the Yorkshire Museum
Museum Gardens, York, YO1 2DR, United Kingdom

Registered Museum No. 1061.

Registered Charity No. 529710.

The cost of marketing this publication was met by an ACCESS grant
from Yorkshire and Humberside Museums Council.

Printed in 10 on 11 point Palatino Typeface
from Author's Disk
by William Sessions Limited
The Ebor Press
York, England

CONTENTS

FOREWORD by Bill Oddie

The study of birds involves more than looking at film footage or ticking off rarities in a standard catalogue. The true lover of ornithology needs to study birds at close quarters to see their plumage, markings and racial characters. This can seldom be done from books. While we may deplore the 19th century collecting which gave us many of our great museum collections, we would be unwise to ignore the immense resource for study which they offer to modern ornithology

The Yorkshire Museum's collection has lain substantially unused for many decades. Many ornithologists simply did not know that it existed. This publication - believed to be the first of its type from a non-national UK museum - will enable the serious ornithologist to see at a glance what the Museum contains and then to use its collection for further study.

I hope that Mike Denton's book will be the first of a series of such specialist ornithological catalogues from the country's major museums.

INTRODUCTION

This catalogue is not intended as an aid to identification; its main aim is to convey to the reader the known breeding distribution (along with the number of described races) of each species contained within the collections and to discuss the geographical provenance of the specimens. By so doing the important collections of the Yorkshire Museum can be used as an invaluable tool for the ornithologist or research worker who wishes simply to study a species in its own right or to concentrate on variations within a species. As most of the species are represented by examples from a wide geographical range, any research worker will find the variations and number of races to their advantage.

The modern day collecting of bird specimens is, fortunately, illegal. Nowadays only museums and a small number of private research workers collect in an effort to aid their own studies and to further Man's knowledge. Most, if not all, of these specimens are acquired through natural deaths and, therefore, the modern day researcher cannot be held responsible for making inroads into populations (as was the case at the end of the 19th century).

The collecting of natural history objects was a Victorian pastime and this perverse mentality can be seen in the pages of books written by ornithologists of the period; the work by Henry Seebohm *The Birds of Siberia* (1901) stands out as a typical example. Seebohm's descriptions of the carnage meted out to the bird-life on his expeditions are legendary and the following paragraphs from his book are a reflection of this attitude:

"I let the gulls go by and aimed at the nearest skua...... Fortunately I brought it down...... I missed several birds, but left seven killed and wounded on the field.

"Whilst I was packing the eggs (Grey Plover)*...... the female came within range, and I took up my gun and shot her......*

"The birds (Grey Plover) *first went up to Harvie-Brown and tried to draw him away by flying about and feigning lameness. They then came up to me and did the same. They were so demonstrative that I felt certain of finding the nest, and shot at the female. I then shot the male...... I caught site of a young Grey Plover in down...... Stopping to pick it up, I saw the nest with three eggs ...... Our attempt to hatch the highly incubated eggs, and thus obtain specimens of young in down, was successful.*

1

"I spent an hour watching it (Pechora Pipit)...... *The bird sang while hovering; it afterwards alighted in a tree, and then descended to the ground...... I shot one, and my companion an hour later shot another.*

"During the time that Glinski was with me he skinned more than a thousand birds......"

It must be remembered, however, that collectors such as Seebohm were pioneers in their field and without these people our knowledge and understanding of birds today would be very impoverished.

The practice of collecting natural history objects was dying out by the mid 20th century and in his monumental book *Yorkshire Birds* (1952) Ralph Chislett was particularly scathing of such collections (both skins and eggs). As stated by Chislett *"To kill birds is a crude method of identification to-day. Egg collecting, and the private collections of set specimens or skins, are selfish pursuits"*. It is fortunate that the old adage 'What's hit is history, what's missed is mystery' is a saying that will forever be linked with the past!

The days when ornithologists studied birds down the barrel of a shotgun are gone forever. Today, unless caught for ringing, birds are studied in the field with binoculars and telescopes; modern day optics and Field Guides have made identification in the field an art that our forefathers could only dream about!

Most of the skin and egg collections left from the Victorian era are now housed in museums. This legacy does, however, give the ornithologist the opportunity to study these specimens in intimate detail. So much so that museum ornithologist Lars Svennson has been able to write an *Identification Guide to European Passerines* and ornithologist Kevin Baker has written the *Identification Guide to European non-passerines*. Both guides are aimed at bird ringers and most of the information contained within these pages has been gleaned using the contents of Museum collections. Without these old collections our knowledge of ageing and sexing live birds would still be in its infancy.

Not only are these skins of use to people who wish to age and sex birds, but they can also be a mine of information for county, and even country avifaunas. The identification of specimens was in its infancy when collecting was in vogue, and identification errors were commonplace. The presence of a skin in a collection has, in many cases, helped to verify a past record. Additionally, skin collections at museums have been found to contain specimens which bear hitherto unrecorded information. The skin collection at the Yorkshire Museum is a typical example and contains specimens which have proved not only to be new to their respective counties but also new to their respective countries. Examples found within the collection at the Yorkshire Museum include Europe's first American Kestrel *Falco sparverius* and France's first White-crowned Black Wheatear *Oenanthe leucopyga*. These and other interesting finds are highlighted in the text.

The value of specimens, especially those of rare or little known species, held in perpetuity in museum collections cannot be over emphasized.

HISTORY

The Yorkshire Museum has had a long, but sometimes chequered, history when it comes to the ownership of bird-skin, -mount and -egg collections. During the years 1842 and 1843 the Museum acquired the Rev. F.O. Morris British Birds' Egg Collection and by 1845 a large ornithological collection embracing over a thousand species had been accumulated.

The importance of the F.O. Morris Egg Collection can be seen by the fact that the contents were used for Fawcett's illustrations in Morris's book *A Natural History of the Nests and Eggs of British Birds* (1875). It would appear that Morris was a contradictory character, however, and his persistant dogfights with other Yorkshire naturalists, including Museum personnel, was probably one of the reasons behind his retrieval of some, if not all, of his collections. There is no record of these retrievals in the *Yorkshire Philosophical Society Annual Reports* and the subsequent history of this important collection is uncertain. It is known, however, that the collection was not known to be at the Museum after about 1870.

In the *Y.P.S. Annual Report* for 1846 it was stated that *"...... by far the most important accession which this department* (Ornithological) *of the Museum has EVER received, has been the magnificent donation by Mr William Rudston Read, of his extensive Collection of British Birds, ... The only condition annexed to this liberal gift was, that it should be kept separate from the general ornithological collection of the Society, and distinguished as "The Rudston Collection of British Birds"......"* It goes on to say that *"The Council propose...... to place the "Rudston Collection" in one of the upper rooms of the Museum, at present occupied by a part of the Collection of Antiquities about to be removed to the Hospitium. It will thus be kept conspicuously distinct, and will permit the more effectual and instructive display of the general collection of Birds in the centre room of the Museum"* The Yorkshire Museum now possessed two bird collections - the main in the central room consisting of more than 1000 species and the Rudston Read Collection which was housed in the upper room.

It is known that during the following year (1847) Rudston Read presented the Museum with a collection of British Birds' Eggs that were housed in a cabinet. No indication as to its size has been ascertained and the collection is no longer extant at the Museum.

In 1866 the Arthur Strickland Collection of British Birds was presented to the Museum. At the time various conditions were attached to this acquisition but they were never published in the *Y.P.S. Annual Reports.* In a letter of 30th November 1866, Mr Rudston Read, writing to the donor of the Collection, states that "......*they were to be deposited in the Museum in the cases as they are delivered without further classification and that the same shall be called the Strickland Collection of British Birds......*" Despite these statements, however, the Strickland Collection appears to have been split up during the next few years - part going into the Rudston Collection and the duplicates into the general collection.

All was not well, however, and in 1868 Rudston Read was to complain that several birds in his collection were totally destroyed by moths, and demanded the dismissal of the Keeper of the Museum. Apparently, moths had first made entry into the collection in 1864 when an infected White-tailed Sea Eagle had been added and which was found to be swarming with maggots. The mount was removed and destroyed, and in 1865 the collection treated with vapour of benzole. In 1866 moths were again to appear and remedial action was taken. There is no mention of this problem in the *Y.P.S. Annual Report,* and the report for 1869 states that the Rudston and Strickland Collections were "......*in excellent order......*" The following sentence in this report is rather contradictory, however, and states that "*The wooden boxes containing the Strickland Collection were found to be infested with the larvae of a wood eating insect* (Anobium tessellatum), *and the ravages thereby caused threatened much injury to the Collection by rendering the specimens easy of access to moths*". The *Y.P.S. Annual Report* for 1872 indicates that few moths were reported during the year but the report for 1873 states that "......*the Collection of British Birds is in a perfect state of preservation......*" Words to this effect also appeared in the 1875 report.

Despite these glowing statements and the lack of any published information to the contrary, quite the opposite was taking place and the *Y.P.S. Annual Report* for 1885 suddenly indicates that the collection "......*was found to be in very poor condition.*" This was a tragic loss of many very important ornithological specimens. Rudston Read died in August 1886 and in 1887 the Rudston Collection - all terms of the gift forgotten - was broken up.

1876 saw the acquisition of a splendid series of bird skeletons that formerly belonged to the late Mr Thomas Allis. Also during this year the Museum was presented with the skeleton of *Dinornis robustus* (Moa).

At the time this was the largest and most perfect specimen known to science. Details of the Moa skeleton, along with the other skeletal preparations at the Museum, can be found in a later chapter.

Between 1888 and 1899 a new Collection of British Birds was formed of which only 15 species (20 specimens) came from The Rudston Read Collection. Some of these, such as the Great Auk, were birds presented to the Y.P.S. and handed over by Rudston Read under the agreement of 1846.

It was in 1902 that James Backhouse decided to present the Museum with his extensive skin collection. As Honorary Curator of Ornithology at the Museum between 1887 and 1921, Backhouse had used his collection to advantage when working on his book *A Handbook of European Birds, for use of Field Naturalists and Collectors* (1890). In 1943 (just before Backhouse's death) work started on combining his collection with that of F.C. Farman (unfortunately nothing is printed regarding this collection) and this amalgamation forms the backbone of the present collection.

In 1942 the Museum was to obtain the British birds egg collection (including cabinets) of Mr G. Howard and Mr E.M. Rutter. It was these collections which formed the basis for the start of the Museum's egg collection. During 1944 the Museum had the good fortune to become the custodian of the Mr William Cooper Collection of Birds' Eggs and the Y.P.S. *Annual Report* for 1945 states that *"It is proposed to make this fine gift the basis of the systematic collection and to utilize the general collection for public display......"* The accumulation of the Museum's general collection received scant attention in the Y.P.S. *Annual Reports* thereafter, most entries simply indicating the acquisition of a small number of eggs during the respective years. By 1948, however, work on the amalgamation of the collections was complete and the results exhibited in the gallery. Details of the contents of these important collections are to be found in a later chapter.

During the intervening years a number of skins and mounts were purchased or deposited at the Museum. These were all small acquisitions with nothing of great significance being received. Time has taken its toll, though, and the whereabouts of a good number of these specimens is unknown. In 1983 this Victorian and now rather antiquated way of displaying bird-mounts was dismantled when it was found that many specimens had been damaged by insects.

The present bird-skin and -mount collections are in excellent condition and have been curated to modern day standards of nomenclature with a view to making them useful to the ornithologist and research worker.

BIRD SKINS

The skin collection contains a total of 3073 specimens that encompasses 482 species (excluding three known hybrids and a possible hybrid) from 67 families. A large proportion of the specimens within the collection, especially the Passeriformes, had originally been identified to subspecific level. Due to the limited time resources no concerted attempt has been made to maintain this standard. The reader will, however, obtain an indication of the subspecies involved simply by looking at the geographical provenance (if accompanied by data) of the specimen concerned.

It will be found when looking at any original data label attached to specimens that a high proportion of the skins were sexed upon dissection. It is worth bearing in mind, however, that over 25% of immatures can be wrongly sexed when using internal organs for sex determination. Great care must, therefore, be taken when utilizing skins sexed by this method. Similarly, care must be taken if biometrics are to be used from such specimens as it is known that shrinkage can cause up to 3% differences in measurements, and wing length cannot be measured to give the maximum possible length.

One of the major problems when dealing with skins that are, in most cases, about a century old, concerns the place names that may be associated with them. Even if the hand writting is discernible the place name may no longer be in current usage. Not only do place names change, but colloquialisms and abbreviations also cloud the issue. Interpretation of locality data can, therefore, be somewhat problematical. It would be misleading to guess at the locations of these place names and, where an exact geographical area cannot be pinpointed, no assumptions have been made. Although every effort has been made to locate and bring place names into the 20th century this has not always been possible. Where there is ambiguity this has been indicated in the text.

Each skin has been allocated a unique Yorkshire Museum accession number and full details are now accessible on a computerised database.

In the following list the species name is followed by a brief account of its known breeding distribution along with the number of described races.

8

GAVIIFORMES

GAVIIDAE

Gavia stellata (Red-throated Diver)
Northern Holarctic. Monotypic. Five specimens (all in non-breeding plumage). With the exception of a skin that is lacking locality data, the others were acquired in the British Isles. One had been killed by flying against the lighthouse at Spurn, Yorkshire on 24th October 1900 (see *Tachybaptus ruficollis* below) and another was taken from the inland site at Broomfleet, Yorkshire on 21st February 1947.

Gavia arctica (Black-throated Diver)
Northern Holarctic. Three races, although some authors now regard *pacifica* as a separate species. The single adult, which is in full breeding plumage, was acquired at Surrendal, Norway on 24th June 1925; the bird was sexed as male upon dissection.

Gavia immer (Great Northern Diver)
Northern Nearctic and, marginally, Palearctic. Monotypic. The single adult, which has lost most of its breeding plumage, was taken at Stromness, Orkney, Scotland on 30th October 1872.

PODICIPEDIFORMES

PODICIPEDIDAE

Tachybaptus ruficollis (Little Grebe)
Palearctic, Ethiopian and Oriental. At least nine races. The nine specimens originate from the British Isles (5); Ireland (1); Hungary (1); India (1) whilst the other has indecipherable locality data. A winter plumaged individual from Minti Jheel, Madubani, India that shows white secondaries may belong to the race *capensis*. A summer plumaged bird that also shows much white in the secondaries may also belong to this race but no geographical provenance can be ascertained as the label is indecipherable, thereby making subspecific determination impossible. A bird from Spurn, Yorkshire was killed by flying into the lighthouse on 23rd October 1900 (see *Gavia stellata* above).

Podiceps grisegena (Red-necked Grebe)
Holarctic. Two races. The two winter plumaged birds both originate from Colchester; one in January 1860 the other in February 1879.

Podiceps cristatus (Great Crested Grebe)
Palearctic, Oriental, Ethiopian (patchily) and Australasian. Three races.
The single specimen, an adult, is showing the last remnants of breeding
plumage and was taken at Fleet Pond, Hampshire on 20th October 1928;
the bird belongs to the nominate race and was sexed as female upon dissection.

Podiceps auritus (Slavonian Grebe)
Holarctic. Monotypic. The single example is in full breeding plumage
and was acquired in Iceland (no collecting date is indicated).

Podiceps nigricollis (Black-necked Grebe)
Holarctic and Ethiopian (patchily). At least three races. The two specimens, both of which exhibit full breeding plumage, are without collecting data.

PROCELLARIIFORMES

PROCELLARIIDAE

Fulmarus glacialis (Fulmar)
Holarctic. At least two races. A single example of a light phase bird from
Haverfordwest, Dyfed, Wales on 27th December 1949.

Puffinus gravis (Great Shearwater)
Gough Island and the Tristan da Cunha group of islands (South Atlantic
Ocean) and the Falklands (South Atlantic Ocean). Monotypic. The single
specimen, said to be a male upon dissection, was acquired at Nolsoy hin,
Faeroe Islands on 2nd October 1945. This bird is of interest as the species
is but an accidental visitor so far north in the Atlantic.

Puffinus griseus (Sooty Shearwater)
Neotropical (southern South America) and Australasian (New Zealand,
south eastern Australia and Tasmania). Monotypic. Three specimens.
One of the skins is of significance as it verifies the first known occurrence of this species in Britain. The bird in question was shot at the mouth
of the Tees, Yorkshire on a stormy day in mid August 1828. The bird was
originally described as an immature Great Shearwater *P. gravis* and further details can be found in Yarrell's *British Birds* Vol. III (in the mid-
nineteenth century the Sooty Shearwater was considered to be the
immature form of Great Shearwater). Another of the birds, shot in Filey
Bay, Yorkshire on 26th August 1887, is mentioned in an article in the

Naturalist 1887: 354. The other skin was acquired at Nolsoy hin, Faeroe Islands on 2nd October 1945.

Puffinus puffinus (Manx Shearwater)
Western Palearctic (mainly Great Britain), Nearctic (patchily) and Australasian (New Zealand only). Six races, although some authors now regard the five in the Pacific as separate species. The four examples were all obtained in the British Isles: two from Anglesey, Wales, one from Dunnington, York, Yorkshire (an unusual inland record) and one from Eynesbury, Saint Neots, Cambridgeshire. This latter mentioned bird is of interest as it was wearing a British Trust for Ornithology ring (number AX7455). An accompanying letter from the BTO indicates that it was one of a number of shearwaters being used for homing experiments by a Dr Matthews of Cambridge University. The birds involved had originally been taken from their breeding-grounds at Skokholm, Pembrokeshire, Wales and released at various points (chiefly inland) after which a watch was kept for their return to the nesting burrow. The release point was not indicated at the time but it was stated that this bird had probably been released at Cambridge.

Puffinus assimilis (Little Shearwater)
Western Palearctic (Azores, Maderia, Cape Verde Islands and Canary Isles), Neotropical (southern South America) and Australasian (western Australia and New Zealand). At least ten races. The single specimen, of the race *baroli,* was taken at Santa Ursula, Tenerife, Canary Isles on 24th May 1889. A note on a reference card for this species indicates that the specimen is probably this species but the wing length (150mm) lies outside the range given by Witherby et al in *The Handbook of British Birds* (1941). Closer inspection, however, reveals that the outer primaries are in moult, thereby accounting for the short wing measurement.

HYDROBATIDAE

Oceanites oceanicus (Wilson's Petrel)
Antarctica north to the subantarctic convergence. Three races. The single specimen is simply labelled "Chatham Moss, 8th October 1886". Unfortunately, no country is indicated and, despite extensive investigations, the country of origin remains unknown. Even though the locality name suggests a British origin, it cannot be assumed that it originated from these islands, and the specimens' provenance may remain a mystery forever! The wing length (155mm) suggests that it may belong to

the race *exasperatus* from the South Shetland Islands and the Antarctic continent.

Hydrobates pelagicus (Storm Petrel)

Western Palearctic (mainly Great Britain and Mediterranean islands). Monotypic. The three specimens originate from British waters: Seaton Carew, Cleveland on 12th Novenber 1868; Billingham, Isle of Wight on 4th April 1948 and Burnham-on-Sea, Somerset on 3rd November 1952.

Oceanodroma leucorhoa (Leach's Petrel)

Northern Atlantic and northern Pacific Oceans. At least four races. Nine specimens. With the exception of two specimens from an untraced locality (both collected on 16th June 1886) the others originate from Greenland (two on an unrecorded date); Clifton Ings, near York, Yorkshire (3rd December 1888); Appleton-le-Moor, Yorkshire (27th December 1927); Wetherby, Yorkshire (30th October 1952); Ingleby Greenhow Moor, Yorkshire (1st November 1952) and Burnham-on-Sea, Somerset (3rd November 1952). The 1952 Yorkshire birds had been picked up either dead or dying after a series of gales that had been the cause of the 'wreck' of this oceanic species.

PELECANIFORMES

SULIDAE

Morus bassanus (Gannet)

Seaboards on both sides of the North Atlantic. Monotypic. The three specimens originate from the east coast of Yorkshire and are represented by a juvenile and two adults. One of the adults is labelled as being caught off Scarborough with a fish-hook on 28th December 1882.

PHALACROCORACIDAE

Phalacrocorax carbo (Cormorant)

Almost cosmopolitan, though not found in the Neotropics. At least five races. The three specimens were all taken at British localities during December (two first-winters in 1887 and 1948 and a second-winter in 1948) and belong to the nominate race.

Phalacrocorax aristotelis (Shag)

Western Palearctic. Three races. The single adult is in full breeding plumage but bears no collecting data.

Phalacrocorax pygmeus (Pygmy Cormorant)
Patchily distributed in the western (southeastern only) and southern central Palearctic. Monotypic. Two specimens. An adult from Slavonia, Turkey on 28th May (no year is indicated) is in full breeding plumage, whilst the other is showing signs of breeding plumage but bears no collecting data.

CICONIIFORMES

ARADEIDAE

Botaurus stellaris (Bittern)
Palearctic and Ethiopian (southern Africa only). Two races. Seven specimens (two pullus and five full-grown). The five full-grown birds belong to the nominate race and, those with collecting data (3), originate from: Aldeburgh, Suffolk on 5th December 1925, Ingmanthorpe, near Wetherby, Yorkshire on 25th February 1947 and the Isle of Wight on an unrecorded date. The Yorkshire bird was found dead under a hedge (an out of context habitat for this reed bed species). The two downy chicks (pullus) have no more precise locality data than "south Russia".

Ixobrychus minutus (Little Bittern)
Western and central Palearctic, Ethiopian, northern Oriental and Australasian. Five races. The single female bears no more precise collecting data than "south Russia, May 1885".

Nycticorax nycticorax (Night Heron)
Almost cosmopolitan (save for Australasian). Four races. The single specimen, an immature, bears no more precise collecting data than "southern Russia, 22nd September 1885".

Ardeola ralloides (Squacco Heron)
Western Palearctic and Ethiopian. Monotypic. Three skins, representing three plumage types. A first-winter was collected at Damiette (now Dumyat), Egypt, Africa on 10th December 1879; a bird in breeding plumage bears no collecting locality but was acquired on 28th May 1883 and an adult in winter plumage was taken at an untraced locality on 13th May 1884

Egretta garzetta (Little Egret)
Western Palearctic (southern), Ethiopian (patchily), Oriental and Australasian. Four races. Two specimens. A breeding plumaged adult

has no more precise collecting data than "south Russia"; the bird is of the nominate race. The other bird, again a breeding plumaged adult, was taken at Lenkoran, Azerbaydzhan, Russia on an unrecorded date. The feet of this latter mentioned bird appear black, a characteristic indicative of the races *immaculatus* and *nigripes*. Closer inspection, however, reveals that the feet have darkened with age and, because of other features, the bird can be assigned to the nominate race.

Egretta alba (Great White Egret)
Almost cosmopolitan. At least four races. A single specimen, sexed as female upon dissection, was collected at Lenkoran, Azerbaydzhan, Russia in April 1910 and belongs to the nominate race.

Ardea purpurea (Purple Heron)
Palearctic, Oriental and Ethiopian. Four races. The three specimens, which all belong to the nominate race, represent a single juvenile, an adult in summer plumage and an adult in winter plumage. The summer plumaged bird bears no collecting data, whilst the winter plumaged bird originated from the south of France on 2nd August 1886. The juvenile was collected in Copenhagen, Denmark but no data is appended.

Ardea cinerea (Grey Heron)
Palearctic, Oriental and Ethiopian. At least four races. Four skins representing pullus (1), immature (2) and adult plumage (1). The adult bird, which is in winter plumage, was acquired at Salton, Yorkshire towards the end of February 1947. One of the immatures had been found ailing on Commercial Street, Norton, Yorkshire and taken into safe keeping but died a few hours later (during the night of 16/17th January 1963); the other immature carries no collecting data. The downy chick (pullus) originates from Cumberland on 6th April 1937.

THRESKIORNITHIDAE

Plegadis falcinellus (Glossy Ibis)
Almost cosmopolitan, though patchily distributed and absent from Neotropics. At least two races. The single specimen, an adult in breeding plumage, was acquired in Yugoslavia on 28th May 1883.

Platalea leucorodia (Spoonbill)
Southern Palearctic, Oriental and Ethiopian (north eastern only). At least three races. A single half grown pullus (the primaries project about 40mm from the sheath) collected in Dobrudshca, Turkey on 5th June 1885 is the only representative of this species.

14

Platalea flavipes (Yellow-billed Spoonbill)
Endemic to Australia and Tasmania. Monotypic. The head and bill of an adult in non-breeding plumage is all that remains of a specimen that carries no data.

ANSERIFORMES

ANATIDAE

Dendrocygna javanica (Indian Whistling Duck)
Oriental. Monotypic. The single specimen, a downy duckling collected in India, had been purchased in 1882.

Cygnus olor (Mute Swan)
Palearctic (introduced in Nearctic). Monotypic. The single adult is devoid of collecting data.

Anser erythropus (Lesser White-fronted Goose)
Extreme northern Palearctic. Monotypic. The single adult was collected in northern Sweden on 10th June 1862.

Anser anser (Greylag Goose)
Palearctic. Two races. The single specimen, a downy gosling, was collected in Romania on 10th May 1874.

Branta bernicla (Brent Goose)
Northern Holarctic. At least three races. The single adult, which belongs to the nominate race, was taken at Colchester, Essex in January 1962.

Alopochen aegytiacus (Egyptian Goose)
Ethiopian (introduced to England). Monotypic. The single specimen, a downy gosling, carries no collecting data.

Tadorna tadorna (Shelduck)
Northwestern and southern Palearctic. Monotypic. The single specimen was collected at Spurn, Yorkshire in January 1909.

Anas penelope (Wigeon)
Northern Palearctic. Monotypic. The two specimens, both males, were taken on the River Swale, Yorkshire in 1968 and at Dobrudscha, Turkey on an unrecorded date.

Anas americana (American Wigeon)
Nearctic. Monotypic. The single female, which bears no data, has a clipped right wing and has obviously originated from captive stock.

Anas formosa (Baikal Teal)
Eastern Palearctic. Monotypic. The two specimens, both males, were acquired at Foochow, China (one on 15th December 1884 whilst the other is devoid of a date).

Anas crecca (Teal)
Holarctic. Two, perhaps 3 races. The single male and female specimens were acquired at Easington, Yorkshire in October 1887. The male is not in full dress and lacks the white scapular stripe of the race *crecca*. The bird can be ascribed to this race, however, on account of the conspicuous buff border encircling the green head band.

Anas platyrhynchos (Mallard)
Holarctic. Seven races. Of the six specimens those with data (4) originate from British localities. Two of the males and one of the females show signs of domesticity.

Anas acuta (Pintail)
Holarctic with isolated populations in the southern Indian Ocean. Three races. The two specimens, both males, were taken at Riccall, Yorkshire in 1900 and Aldeburgh, Suffolk on 15th December 1924.

Anas querquedula (Garganey)
Palearctic. Monotypic. The two specimens, both males, were acquired in Lincolnshire in March 1873 and Lancashire on 25th April 1891.

Anas clypeata (Shoveler)
Holarctic. Monotypic. The single specimen, a female, was collected at Spurn, Yorkshire in September 1886.

Marmaronetta angustirostris (Marbled Teal)
Patchily distributed between Spain in the west and northwestern India in the east. Monotypic. The single specimen, sexed as male upon dissection, was acquired at Lenkoran, Azerbaydzhan, Russia on an unrecorded date.

Netta rufina (Red-crested Pochard)
Western (patchily) and central Palearctic. Monotypic. The two male specimens are both in full summer plumage. One bird carries no collecting data whilst the other was taken at Lenkoran, Azerbaydzhan, Russia on an unrecorded date.

Aythya valisineria (Canvasback)
Nearctic. Monotypic. The single specimen, a downy duckling, carries no collecting data.

Aythya ferina (Pochard)
Palearctic. Monotypic. The single male was taken near Hull, Yorkshire on 29th December 1944.

Aythya nyroca (Ferruginous Duck)
Western (patchily) and central Palearctic. Monotypic. The single specimen has indecipherable locality data but was sexed as female upon dissection.

Aythya fuligula (Tufted Duck)
Palearctic. Monotypic. The single male was collected near Uffington, Shropshire on 17th February 1925.

Aythya marila (Scaup)
Northern Holarctic. At least two races. Single male and female specimens. The male, a first-year bird, was taken at Colchester, Essex in January 1861. The female, again a first-year bird, was acquired at Aldeburgh, Suffolk on 5th December 1925.

Somateria mollissima (Eider)
Northern Holarctic. Six races. Five specimens (a downy duckling, three adult males and a female). The males were collected in Scotland (29th May 1891 and 5th June 1891) and the Faeroe Islands (9th July 1946), the female originates from Shetland, Scotland (2nd October 1937) whilst the duckling was taken at Spitzbergen on an unrecorded date.

Somateria spectabilis (King Eider)
Extreme northern Holarctic. Monotypic. Single male and female specimens. The adult male is simply labelled "Arctic Regions" whilst the adult female was collected in the Davis Strait (between Baffin Island and Greenland) in June 1876.

Somateria fischeri (Spectacled Eider)
Eastern Palearctic (extreme north) and western Nearctic. Monotypic. The single specimen, a downy duckling, has no more precise collecting data than "north Russia, 25th July 1875".

Histrionicus histrionicus (Harlequin)
Eastern Palearctic and east and west Nearctic. Monotypic. Four specimens; an adult male from Labrador, Canada in 1878 and three female-types from the Davis Strait (between Baffin Island and Greenland) in June 1876, Iceland on an unrecorded date and one that is devoid of collecting data.

Clangula hyemalis (Long-tailed Duck)
Holarctic. Monotypic. Nine specimens. Due to the complicated moult sequence exhibited by this species identification other than adult males is difficult and no attempt has been made to sex other than birds so plumaged. The three adult males were acquired in Iceland (9th June 1875); Rugen, Germany (4th May 1883) and Labrador, Canada on an unrecorded date. The others originate from Iceland (five specimens taken on 16th September 1884 including a well grown duckling) and Easington, Yorkshire on 11th October 1887.

Melanitta nigra (Common Scoter)
Northern Palearctic and northern Nearctic (patchily). Two races. Single male and female specimens. The adult male, which belongs to the nominate race, was shot at Wintersett Reservoir, near Wakefield, Yorkshire on 8th January 1947. The adult female was taken at Surrendal, Norway on 19th September 1924.

Melanitta fusca (Velvet Scoter)
Northern Holarctic. At least three races. The two males, which belong to the nominate race, were both acquired at Stromness, Orkney, Scotland; a first-year on 8th April 1938 and an adult on 30th January 1939.

Bucephala clangula (Goldeneye)
Holarctic. Two races. Three specimens (an adult male and two adult females). One of the females was acquired at Colchester, Essex in January 1862 whilst the other is devoid of locality data but was collected in 1897; the male was taken at Fort William, Highland, Scotland on 5th November 1924.

Mergus serrator (Red-breasted Marganser)
Northern Holarctic. Monotypic. Three specimens; two adult males and a first-winter (sexed as female upon dissection). The males originate from North Wales on 20th December 1925 and North Uist, Scotland on 16th January 1926 whilst the first-year was collected at Surrendal, Norway on 16th October 1924.

Mergus merganser (Goosander)
Holarctic. Three races. Four of the five specimens (three adult males and a first-year female) were collected from Yorkshire waters during the winter months. The other, an adult male, was taken near Brampton, Cumberland on 23rd December 1928.

FALCONIFORMES

PANDIONIDAE

Pandion haliaetus (Osprey)
Cosmopolitan, though patchy in Ethiopian and Neotropics. Five or six described races. Two specimens; an adult male from Brandenburg, Germany taken on 25th July 1876 and an adult female shot at Hornsea, Yorkshire on 24th May 1943.

ACCIPITRIDAE

Pernis apivorus (Honey Buzzard)
Western Palearctic. Monotypic. Of the six specimens five bear data; Sweden on 20th June 1885; Spurn, Yorkshire in October (no year appended); one has no more precise locality data than "southern Russia, May 1884" and the other two, although collected on 2nd May 1885 and 10th July 1878, are from an untraced locality.

Elanus caeruleus (Black-shouldered Kite)
Southwestern Palearctic (patchily), Oriental and Ethiopian. At least four races. The single adult was collected in Morocco, Africa on 14th April 1886.

Milvus migrans (Black Kite)
Palearctic, Ethiopian, Oriental and Australasian. At least six races. Of the four specimens only two carry collecting data; south Lapland (country not stated) on 9th August 1885 and an untraced locality on an unrecorded date.

Milvus milvus (Red Kite)
Western Palearctic. Two races. Three of the four specimens bear no collecting data but were taken on the following dates; 10th October 1876 (first-winter), 22nd September 1885 (first-winter) and 19th June 1886 (juvenile). The only individual with full data, an adult, was found dead near Scagglethorpe, Yorkshire on 13th April 1976.

Haliaeetus albicilla (White-tailed Sea Eagle)
Palearctic and Nearctic (southwest Greenland). Monotypic. Single immature and adult specimens. The immature was taken at Nordmore, Norway on 19th March 1926 and the adult was acquired at the same locality some six months later.

Circaetus gallicus (Short-toed Eagle)
Western and central Palearctic and Oriental. Probably monotypic, although some authors describe up to four races. The single specimen bears no collecting data.

Circus aeruginosus (Marsh Harrier)
Palearctic, Ethiopian (Madagascar only) and Australasian. Nine races. Four specimens. With the exception of individuals from Sarepta, Kalmytskaya, Russia in May 1883 and south Lapland (country not indicated) on 9th September 1885 the other specimens bear no collecting data, although one was acquired on 2nd October 1883.

Circus cyaneus (Hen Harrier)
Holarctic. Four races. The five specimens represent three females and two second-summer males. A male and female were collected in south Lapland (country not indicated) on 30th September 1885 and 22nd April 1886 respectively. One of the other females was acquired at Crockey Hill, York, Yorkshire in 1951. The other specimens bear no data although the male was collected on 10th May 1885 and the female on 2nd February 1886.

Circus macrourus (Pallid Harrier)
Central Palearctic. Monotypic. Four skins representing adult male (1), second-summer male (2) and female-type (1) plumages. Two of the specimens bear no more precise collecting data than "southern Lapland, 10th May 1885" (adult male) and "southern Russia, May 1885" (second-summer male). The other second-summer male was acquired at an untraced locality on 10th May 1877 whilst the female-type is devoid of data.

Circus pygargus (Montagu's Harrier)
Western and central Palearctic. Monotypic. The single specimen, which is in female-type plumage, was taken in Andalusia, Spain on an unrecorded date. Additionally, there is the right wing of an immature that was shot above Pateley Bridge, Yorkshire in August 1951.

Accipiter gentilis (Goshawk)
Holarctic. Between seven and nine races described, although some authors regard the North American races as a separate species (*Accipiter atricapillus*). The three specimens represent an adult and two immatures. The adult was taken in Lapland (country not indicated) on 7th May 1877; one of the immatures was acquired in Copenhagen, Denmark on an unrecorded date whilst the other carries no data.

Accipiter nisus (Sparrowhawk)
Palearctic. Six races. The 31 specimens represent all age groups and both sexes. All the birds with locality data (29) originate from sites in the British Isles. An adult female, along with her three young, was taken at a nest in High Force Wood, Teesdale, Yorkshire on 25th June 1886; the male, which very rarely visits the nest in this species, was not shot and lived to tell the tale! Another adult female had been snared in a pole trap set on Askrigg Moor, Yorkshire in August 1900.

Buteo buteo (Buzzard)
Palearctic. Up to 16 races described. Of the ten specimens only four carry locality data; Highland, Scotland (1); Caernarvonshire, Wales (2) and a single from the Middle Volga, Kalmytskaya, Russia on 13th March 1884.

Buteo lagopus (Rough-legged Buzzard)
Northern Holarctic. Four or five races. Five specimens. A single was collected at Hammerfest, Norway on 2nd July 1888 and another was found dead in Forge Valley, Scarborough, Yorkshire on 22nd November 1947. This latter bird had been present in the area for several days before its demise. The other three specimens (acquired on 8th January 1878, 6th April 1883 and 7th October 1883) originate from untraced localities.

Buteo rufinus (Long-legged Buzzard)
Western (mainly north Africa) and central Palearctic. Two races. The single specimen is devoid of collecting data.

Aquila clanga (Great Spotted Eagle)
Palearctic. Monotypic. Two adult specimens. A bird from the Middle
Volga, Kalmytskaya, Russia was collected on 20th February but no year
is given, whilst the other carries no collecting data.

Hieraaetus fasciatus (Bonelli's Eagle)
Western Palearctic (southern), Ethiopian and Oriental. Three races. The
single immature was collected in Sardinia on an unrecorded date.

Hieraaetus pennatus (Booted Eagle)
Palearctic (patchily) and Ethiopian (south Africa only). Monotypic. The
two specimens, which represent both dark and light phases, originate
from the south of Spain (dark phase) on 17th September 1884 and Spain
(light phase) on an unrecorded date.

FALCONIDAE

Falco sparverius (American Kestrel)
Nearctic and Neotropical. 14 races. The single specimen, an adult female,
bears a label indicating that it was acquired at Helmsley, Yorkshire in
May 1882. The species was added to the Western Palearctic List in 1901
(Denmark) and there have been two records for the British Isles (Fair
Isle, Shetland and Cornwall, both in 1976). In a situation where a skin
has been found bearing information of this nature the possibility of an
escape from captivity or fraud are underlying factors and cannot always
be ruled out. Several factors, however, are in favour of this individual
being a genuine vagrant:
1) Investigations at the Natural History Museum, Tring have revealed
   that the specimen belongs to either the nominate race or the race
   phalaena. These are North American races and, therefore, the ones
   most likely to make a transatlantic crossing.
2) The locality data, along with the correct species name, is contained
   on the single label, indicating that no transposing of labels has taken
   place.
3) The specimen's claws and upper mandible are not overgrown or
   damaged (this is a common problem with birds that have been kept
   in captivity).
4) The plumage is in pristine condition and shows no feather damage
   (broken feathers or worn plumage, especially tail feathers, are a sign
   of captivity).
5) The record would have been made known to the ornithological world
   had 'fame' been the main reason behind a fraud. The procurer of the
   specimen, although adding the correct species name, had made no

attempt to project this as the first European record and the possibility of fraud is therefore diminished.
This hitherto unpublished record is of the greatest significance. A full description along with photographs has been forwarded to the British Ornithologists' Union Records Committee and their deliberations on this potential first for the Western Palearctic are awaited.

Falco tinnunculus (Kestrel)
Palearctic, Oriental and Ethiopian. At least 11 races. The 18 skins represent all age groups and both sexes. Of the 14 specimens with data ten originate from the British Isles. The other four are from Nolsoy hin, Faeroe Islands on 11th October 1946; southern India on an unrecorded date whilst the other two are from untraced foreign localities. The Indian bird has leather jessies attached to its legs and is therefore an ex-falconer's bird.

Falco vespertinus (Red-footed Falcon)
Central and eastern Palearctic. Monotypic. The single specimen, an adult male, bears no more precise collecting data than "south Russia, 4th April 1885".

Falco columbarius (Merlin)
Holarctic. 11 races. Seven specimens. The specimens with locality data (5) originate from Spurn, Yorkshire (2); Caithness, Scotland (1); Yaynol, Caernarvonshire, Wales (1); Surrendal, Norway (1) and aboard the S.S. Thyra when she was on the Island Voyage in the North Sea on 30th August 1884.

Falco subbuteo (Hobby)
Palearctic and, marginally, Oriental. Two races. The three specimens originate from the south of France (juvenile) in 1885; Sarepta, Kalmytskaya, Russia (adult) on 28th April 1886 and Easington, Yorkshire (juvenile) on 1st September (no year appended).

Falco eleonorae (Eleonora's Falcon)
Western Palearctic (mainly Mediterranean islands). Monotypic. The single juvenile was taken at an untraced locality on 6th November 1885.

Falco rusticolus (Gyrfalcon)
Holarctic. Monotypic. Of the eight specimens six bear collecting data; Greenland (3), Iceland (1), Norway (1) and Denmark (1). This last mentioned bird is of interest as the species is but an accidental visitor to

23

Denmark. The bird in question was collected in Copenhagen on an unrecorded date.

Falco peregrinus (Peregrine)
Cosmopolitan. At least 15 races. The eight skins represent both sexes and all plumage types with the exception of adult female. Five of the specimens (adult male, immature male and three immature females) were collected in the British Isles. The other three are from Nordmore, Norway (immature male), southern Spain (immature male) and south Russia (immature female).

GALLIFORMES

TETRAONIDAE

Lagopus lagopus (Willow/Red Grouse)
Holarctic. At least 17 races. Of the eight specimens five belong to the race *scoticus* and, with the exception of a bird devoid of collecting data, were acquired on Yorkshire moors. A bird of the nominate race was taken at Karesuando, Sweden on 6th October 1884 and is attaining winter plumage. The other two specimens belong to one of the races which acquires a white winter plumage and were purchased at York Market, Yorkshire in December 1885. The white-winged races are not native to the British Isles and these birds must have been imported.

Lagopus mutus (Ptarmigan)
Holarctic. At least 23 races. The 21 skins represent the summer and winter plumages of both sexes. With the exception of a female from Greenland (no locality indicated) on 17th June 1879 and a male from Scandinavia (country not indicated) on 22nd January 1885, the others originate from Iceland (mainly during September 1884).

Tetrao tetrix (Black Grouse)
Palearctic. Up to eight races described. The two specimens, a male and a female, were acquired in Teesdale, Yorkshire in December 1886.

Tetrao urogallus (Capercaillie)
Western and central Palearctic. Up to ten races described. Three specimens (two adult males and a downy chick). One of the males had been brought from Norway and sold in the market at York, Yorkshire in March 1888 and the other had been kept in confinement at Scampston, Yorkshire before dying in November 1909. This was the easy way to acquire

specimens! The downy chick, which has flight feathers protruding about 50mm from the sheath, was collected at Pitlochry, Tayside, Scotland on 15th June 1905.

Bonasa bosania (Hazel Grouse)
Palearctic. At least six races. Five specimens. Single males were collected in Sweden and Norway. The others, two males and a female, originate from different untraced localities.

PHASIANIDAE

Alectoris chukar (Chukar)
Southern Palearctic. Introduced to the western Nearctic. Up to 17 races described. The single specimen originates from an untraced locality and was collected on 2nd September 1885.

Alectoris barbara (Barbary Partridge)
Southwestern Palearctic (mainly north Africa). Four or five races. The single specimen was shot at Tangier, Morocco, Africa on 14th January 1870.

Alectoris rufa (Red-legged Partridge)
Southwestern Palearctic (introduced to Great Britain). Up to five races described. The two specimens originate from Lincolnshire on 26th October 1882 and Hyeres, France on 19th December 1890.

Perdix perdix (Grey Partridge)
Western and central Palearctic. At least nine races. Nine specimens. With the exception of two skins that are devoid of locality data the others were acquired in England (5) and an untraced foreign locality (2). Eight of the specimens are full-grown birds but a pullus that is less than a week old was taken in Yorkshire on 19th August 1882.

Coturnix coturnix (Quail)
Palearctic, Oriental and Ethiopian. Five or six races. The two specimens were both purchased at Ajaccio market, Corsica on 9th January 1891.

Coturnix japonica (Japanese Quail)
Extreme southeastern Paleactic and northern Oriental. Monotypic. The single specimen has a label attached that indicates that it was from an aviary on 18th January 1897 and that it had been sexed as male upon dissection.

Phasianus colchicus (Pheasant)
Central and eastern Palearctic and Oriental (introduced to North America and Europe, including Great Britain). Up to 34 races described. The four adult males originate from British localities, each bird exhibiting a different assortment of colorations.

Chrysolophus pictus (Golden Pheasant)
Southeastern Palearctic (indigenous populations only in mountains of central China). Monotypic. An adult male and the tail of a male are both devoid of data.

GRUIFORMES

TURNICIDAE

Turnix sylvatica (Andalusian Hemipode)
Palearctic (patchily in Spain and north Africa), Oriental and Ethiopian. Up to 23 races described. The single specimen was collected in southern Spain on an unrecorded date.

GRUIDAE

Anthropoides virgo (Demoiselle Crane)
Southern Palearctic. Monotypic. The single downy chick bears no more precise locality data than "south Russia".

RALLIDAE

Rallus aquaticus (Water Rail)
Palearctic. Four races. The 15 skins represent all plumage types (with the exception of pullus) and both sexes. The 11 specimens with locality data originate from Yorkshire (5); Devon (1); Hampshire (1); Pembrokeshire, Wales (1); Islay, Hebrides, Scotland (1) and Corsica (2). The birds from Corsica had been purchased at Ajaccio market on 9th January 1891 and 27th December 1891.

Gallirallus australis (Weka)
Endemic to New Zealand. Four races. A single , which is probably of the nominate race, was taken on South Island, New Zealand on an unrecorded date.

Crex crex (Corncrake)
Palearctic. Monotypic. The seven specimens were acquired at English localities; Durham (July 1880); Yorkshire (May 1889 and September 1950); Dorset (September 1947); Gloucestershire (April 1949); Cornwall (May 1949) and Huntingdonshire (August 1952). The 1889 Yorkshire bird had been picked up in the North Sea off Spurn.

Porzana parva (Little Crake)
Western and central Palearctic. Monotypic. Four specimens. An adult male and an adult female were taken in southern France and a juvenile and an adult female originate from untraced localities.

Porzana pusilla (Baillon's Crake)
Palearctic, Ethiopian and Australasian. Six or seven races. Four specimens. With the exception of an adult from an untraced locality the others were acquired in southern France (adult); Majorca, Balearic Isles (adult) and Calcutta, India (juvenile).

Porzana porzana (Spotted Crake)
Western and central Palearctic. Monotypic. Ten specimens. With the exception of a specimen that is devoid of locality data the others originate from England (3); France (2); Malta (1); Switzerland (2) and Russia (1).

Porzana carolina (Sora)
Nearctic. Monotypic. Two juvenile specimens. One of the specimens was collected in Scarboro, New Hampshire, USA on 9th September 1886. The other is labelled as being acquired in Switzerland during 1877. A label, attached to this specimen by the late R.Wagstaffe, indicates that the collecting data is dubious. I can do no more than reflect this sentiment. It is conceivable, however, that there is an untraced locality of that name within the species' range (see *Dryocopus pileatus* below).

Gallinula chloropus (Moorhen)
Almost cosmopolitan (save for Australasian). Up to 12 races described. The seven specimens, which represent the plumages of juvenile, first-year and adult, were collected at British localities.

Porphyrio porphyrio (Purple Gallinule)
Western Palearctic (patchily), Oriental, Ethiopian and Australasian. At least 16 races. The single adult, which belongs to the race *madagas-*

cariensis, was collected at Damiette (now Dumyat), Egypt, Africa on the 12th July of an unrecorded year.

Fulica atra (Coot)
Palearctic, Oriental and Ethiopian. Four races. Three specimens (a pullus and two full-grown). The two full-grown birds were taken at Croxdale, Durham on 5th April 1875 and Hull, Yorkshire on 25th January 1947 whilst the downy chick (pullus) originates from Cropstone Reservoir, Leicestershire on 4th May 1922.

Fulica cristata (Crested Coot)
Southwestern Palearctic (patchily in southern Spain and Morocco) and Ethiopian. Monotypic. The single specimen was collected in southern Spain on the 30th June of an unrecorded year.

OTIDAE

Tetrax tetrax (Little Bustard)
Southwestern and central Palearctic. Probably monotypic, although some authors recognize two races. The single specimen, in female-type plumage, was taken at an untraced locality.

Otis tarda (Great Bustard)
Palearctic (patchily). Two, although some authors recognize three, races. The single adult male, which exhibits full breeding plumage, carries no collecting data.

CHARADRIIFORMES

HAEMATOPODIDAE

Haematopus ostralegus (Oystercatcher)
Palearctic. Three races. The 14 skins represent all plumage types (with the exception of pullus). Of the nine specimens with locality data, six originate from the British Isles and three from the Faeroe Islands.

RECURVIROSTRIDAE

Recurvirostra avosetta (Avocet)
Palearctic and Ethiopian. Monotypic. The single specimen, a first-year, is devoid of collecting data.

BURHINIDAE

Burhinus oedicnemus (Stone-curlew)
Southwestern Palearctic and Oriental. Up to eight races described. The four specimens originate from Perpignan, France (bought at the local market in January 1886); Winterton, Norfolk (5th April 1886 and 6th April 1886) and Wetwang, Yorkshire (July 1909).

GLAREOLIDAE

Pluvianus aegyptius (Egyptian Plover)
Ethiopian (patchily). Probably monotypic, although some authors recognize two races. The single specimen, an adult, was taken on the River Benue at Bagana, Nigeria, Africa on 6th January 1908.

Glareola pratincola (Collared Pratincole)
Southern Palearctic and Ethiopian (patchily). Up to five races described. The single specimen, a juvenile, was collected at Dobrudscha, Turkey on 9th September 1885.

CHARADRIIDAE

Vanellus vanellus (Lapwing)
Palearctic. Monotypic. The five skins, representing first-year and adult plumages, were collected in Yorkshire with the exception of a first-year from Mersea, Essex.

Pluvialis apricaria (Golden Plover)
Northwestern Palearctic. Monotypic. Two races formerly described but now found to be clinal and not to warrant subspeciation. 14 skins, representing both summer and winter plumages. Of the 13 specimens with data, ten were collected at British localities whilst the others originate from Iceland on 10th July 1874 and Nolsoy hin, Faeroe Islands on 10th October 1946 and 16th November 1946.

Pluvialis fulva (Pacific Golden Plover)
Northern Palearctic and extreme northwestern Nearctic. Monotypic. Two specimens. One of the specimens was taken at Foochow, China in April 1884 whilst the other, a juvenile, bears no locality data but was collected on 15th September 1886.

Pluvialis squatarola (Grey Plover)
Northern Holarctic. Monotypic. The five specimens represent first-year and winter adult plumages. With the exception of a first-year that is

29

devoid of collecting data the others originate from Northumberland (first-year), Yorkshire (two first-years) and Lancashire (adult winter).

Charadrius hiaticula (Ringed Plover)
Northern Palearctic and northeastern Nearctic. Two races. The 19 skins represent pullus, juvenile, first-year and adult plumages. Of the 18 specimens with data, 16 originate from the British Isles (including the pullus) whilst the others were acquired at Hammerfest, Norway and along the Yushina River, Russia.

Charadrius dubius (Little Ringed Plover)
Palearctic, Oriental and northern Australasian. Three races. The two specimens, both adults, originate from Malta on 18th May 1861 and the Punjab, India on an unrecorded date.

Charadrius alexandrinus (Kentish Plover)
Cosmopolitan (save for Australasian). Six races. The five specimens were acquired in France (female); Morocco, Africa (male and female); northern India (juvenile) and China (male).

Charadrius asiaticus (Caspian Plover)
Central Palearctic. Monotypic. Three specimens. A bird in winter plumage was taken in north Africa on 13th February 1864; a summer plumaged bird was acquired in South Africa in February 1896 and a juvenile was collected in Malta on an unrecorded date. This last mentioned bird is of interest as the species is but an accidental visitor to Malta.

Charadrius morinellus (Dotterel)
Palearctic (mainly northern). Monotypic. Three specimens; two of which were taken on spring migration in Yorkshire, the other in mid-winter in Egypt, Africa.

SCOLOPACIDAE

Limosa limosa (Black-tailed Godwit)
Palearctic. Three races. The single specimen, which exhibits full summer plumage, was acquired in Lincolnshire on 30th March 1908.

Limosa lapponica (Bar-tailed Godwit)
Northern Palearctic and northwestern Nearctic. Two races. The six specimens were taken on Holy Island, Northumberland (two juvenile males and an unsexable juvenile); Blakeney, Norfolk (unsexable juvenile);

Withernsea, Yorkshire (unageable female) and Aldeburgh, Suffolk (adult female).

Numenius phaeopus (Whimbrel)
Northern Holarctic. Four races. The six specimens were collected on the Isle of Wight on 15th May 1886 (1) and Nolsoy hin, Faeroe Islands on 28th May 1946 (4) and 1st June 1946 (1).

Numenius arquata (Curlew)
Palearctic. Two races. The two specimens were acquired at Spurn, Yorkshire; one on 4th September 1882 the other on 28th August 1886.

Tringa erythropus (Spotted Redshank)
Northern Palearctic. Monotypic. The single specimen, a bird in winter plumage, was acquired in Lincolnshire in September 1901.

Tringa totanus (Redshank)
Palearctic. Up tp six races described. The single specimen, which bears no locality data, was collected in 1949.

Tringa stagnatalis (Marsh Sandpiper)
Central and eastern Palearctic. Monotypic. Two specimens. A specimen that bears no more precise collecting data than "south Russia, 11th May 1885" is in full summer plumage. The other, a winter plumaged bird, was acquired at Butiaba, Lake Albert, Uganda, Africa on 16th November 1901.

Tringa nebularia (Greenshank)
Palearctic. Monotypic. The three specimens were acquired in Highland, Scotland (an adult in summer plumage) on 15th July 1885; Spurn, Yorkshire on 22nd August 1885 and Annan, Dumfries and Galloway, Scotland on 19th May of an unrecorded year.

Tringa ochropus (Green Sandpiper)
Palearctic. Monotypic. The four specimens were collected at Perpignan, France (bought at the local market) on 5th January 1886; Harome, Yorkshire in July 1888; Ubekskaya, Russia on 18th September 1929 and Bentley, Leicestershire on 25th July 1934.

Tringa glareola (Wood Sandpiper)
Palearctic. Monotypic. The five specimens were acquired in Norway (5th June 1882); Britain (22nd June 1883); Finland (no more precise date than

1900); Switzerland (8th April but no year appended) and Malta (6th May but no year appended).

Xenus cinereus (Terek Sandpiper)
Central and eastern Palearctic. Monotypic. Two specimens; an adult in summer plumage from Arkhangel'sk, Russia on 25th May 1885 and a winter plumaged bird from Foochow, China on 18th October 1886.

Actitis hypoleucos (Common Sandpiper)
Palearctic. Monotypic. Nine specimens (a pullus and eight full-grown). With the exception of full-grown birds from Norway (2) and Morocco, Africa (1) the others were taken in England (four in Yorkshire and one in Lancashire). The bird from Morocco, collected at Tangier on 25th April 1902, is showing albinistic tendencies. The bird in question exhibits white greater coverts and fourth primary (numbered ascendantly) on both wings. The downy chick (pullus) originates from Cumberland in the June of an indecipherable year.

Actitis macularia (Spotted Sandpiper)
Nearctic. Monotypic. The single specimen, an adult with remnants of summer plumage, was taken in Costa Rica, Central America on 24th October 1897.

Arenaria interpres (Turnstone)
Northern Holarctic. Two races. The 12 skins represent the plumages of juvenile and adults in both winter and summer dress. With the exception of a specimen that is devoid of collecting data the others originate from Nolsoy hin, Faeroe Islands (four taken on 6th September 1946) and Spurn, Yorkshire.

Phalaropus lobatus (Red-necked Phalarope)
Northern Holarctic. Monotypic. The five specimens, which are all in summer plumage, were taken at Alexievka, Russia on 9th July 1875; Lapland (country not indicated) on 13th June 1885; Sweden on 2nd August 1887 and Greenland (2) on unrecorded dates.

Phalaropus fulicarius (Grey Phalarope)
Northern Holarctic. Monotypic. The five skins represent the plumages of first-year (2) and adults in summer plumage (3). The four specimens with data originate from Greenland (adult); Iceland (adult); York, Yorkshire (first-year) and Bridlington, Yorkshire (first-year).

Scolopax rusticola (Woodcock)
Palearctic. Monotypic. Three specimens; a small pullus from Helmsley, Yorkshire on 26th May 1946, a full grown bird from Scauby, Lincolnshire on an unrecorded date and a full grown bird that is devoid of collecting data.

Gallinago stenura (Pin-tailed Snipe)
Eastern Palearctic. Monotypic. The single specimen, which is erroneously labelled *Gallinago media*, was collected in Madras, India on an unrecorded date.

Gallinago media (Great Snipe)
Northwestern Palearctic. Monotypic. The three specimens were collected in northern Russia on 22nd June 1875 (adult), Yorkshire on 4th September 1881 (first-year) and Sweden on 2nd August 1883 (adult). The Yorkshire bird, from Easington, is of interest as the record was not published in Nelson's *The Birds of Yorkshire* (1907) or Chislett's *Yorkshire Birds* (1952). It is known from these works that *G. media* was recorded more frequently at the turn of the century than today. This specimen does, however, add an hitherto unrecorded occurrence to the Yorkshire list.

Gallinago gallinago (Snipe)
Holarctic, Ethiopian and Neotropical. Three races. With the exception of three specimens from Nolsoy hin, Faeroe Islands (28th April 1945, 1st May 1945 and 29th February 1946) the other twelve were collected during the winter months at British or Irish localities. The specimens from the Faeroe Islands exhibit the narrow mantle/scapular stripes indicative of the race *faroensis.*

Lymnocryptes minima (Jack Snipe)
Northern Palearctic. Monotypic. With the exception of a specimen devoid of collecting data the other six were acquired in Yorkshre (3); Leicestershire (1); Norfolk (1) and Caernarvonshire, Wales (1).

Calidris canutus (Knot)
Holarctic. Four races. The 34 skins represent the plumages of juvenile, first-winter and adult (both summer and winter). With the exception of an adult that was found exhausted near Helmsley, Yorkshire on 12th February 1953 all the others were acquired at British coastal localities (mainly Spurn, Yorkshire and the Ribble Estuary, Lancashire).

33

Calidris alba (Sanderling)
Holarctic. Monotypic. The 18 skins represent the plumages of juvenile, first-winter and adult (summer and winter). The specimens with data (15) originate from the British Isles with the exception of a juvenile from north Africa , a summer plumaged adults from Spain and a partial summer plumaged adult from Labrador, Canada.

Caldiris minuta (Little Stint)
Northern Palearctic. Monotypic. The seven specimens, representing both juvenile and adult plumages, originate from the British Isles with the exception of an adult from Germany and two adults which bear no more precise locality data than "south Russia" and "north Russia".

Calidris temminckii (Temminck's Stint)
Northern Palearctic. Monotypic. The two specimens were collected in Egypt, Africa on 13th January 1875 and Norway on 6th June 1898 (an adult in breeding plumage).

Calidris bairdii (Baird's Sandpiper)
Northern Nearctic and northeastern Palearctic. Monotypic. Two specimens. One of the specimens, a juvenile, was taken in Colorado, North America on 9th August 1897 whilst the other, an adult, carries no collecting data.

Calidris melanotos (Pectoral Sandpiper)
Northern Nearctic and northeastern Palearctic. Monotypic. The single specimen, an adult, was acquired on the Isle of Unst, Shetland, Scotland on 20th June 1839.

Calidris maritima (Purple Sandpiper)
Northern Holarctic. Monotypic. Ten specimens. With the exception of an adult in full breeding plumage (no collecting data is attached) all others are British or Faeroe Island birds taken during the winter months.

Calidris alpina (Dunlin)
Holarctic. Six races. The 39 skins represent the plumages of pullus, juvenile, first-winter and adult (both summer and winter). Of the 35 specimens with data, 32 originate from the British Isles, a single was acquired in California, North America and the other two have no more precise locality data than "south Russia". The wide geographical provenance and the differing bill lengths exhibited by the specimens ensures that a good number (if not all) of the described races are available.

Calidris ferruginea (Curlew Sandpiper)
Northeastern Palearctic. Monotypic. The ten specimens were acquired at British localities when on autumn migration and are all in juvenile plumage.

Eurynorhynchus pygmeus (Spoon-billed Sandpiper)
Extreme eastern Palearctic. Monotypic. The single specimen was collected at Foochow, China on 22nd September 1886.

Philomacus pugnax (Ruff)
Northern Palearctic. Monotypic. Nine skins representing the plumages of pullus, adult female, juvenile male and adult male in breeding dress. With the exception of two specimens (adult male and adult female) which are devoid of locality data the others were acquired in Yorkshire (juvenile male); Germany (adult male); Holland (two adult males); Denmark (adult male), an untraced locality in Lapland (country not indicated) (down chick) and Nigeria, Africa (adult female).

STERCORARIIDAE

Stercorarius skua (Great Skua)
Western Palearctic, southern oceans and Antarctic. Up to six races described. The two specimens were both taken on the Faeroe Islands; one in 1884 the other on 10th October 1946.

Stercorarius pomarinus (Pomarine Skua)
Northern Holarctic. Monotypic. Six skins, representing light morph adults and juveniles. The specimens originate from Whitby, Yorkshire on 27th October 1879; Spurn, Yorkshire (no date appended); Lapland (no country indicated) on 9th June 1885 and Nolsoy, Faeroe Islands on 12th October 1946 (2) and 15th November 1946.

Stercorarius parasiticus (Arctic Skua)
Northern Holarctic. Monotypic. 11 skins representing adults of both colour morphs and juveniles. Four specimens originate from Yorkshire coastal localities; Spurn on 1st October 1886; Bridlington in October 1887 (2) and Scarborough on an unrecorded date. The seven others were taken in Iceland on 29th June 1885 (1) and the Faeroe Islands on 4th July 1946 (3) and 8th July 1946 (3).

Stercorarius longicaudus (Long-tailed Skua)
Northern Holarctic. Two races. Two specimens. An adult from Labrador, Canada was acquired in 1884 whist the other, a bird in sub-adult plumage, was picked up dead at Duncombe Park, Helmsley, Yorkshire on 23rd September 1950. This latter bird is of interest as it was found at an inland locality. On 21st October 1950 the bird (then a mounted specimen) was exhibited at a meeting of the Yorkshire Naturalists' Union in Leeds, Yorkshire.

LARIDAE

Pagophila eburnea (Ivory Gull)
Northern Holarctic. Monotypic. The single first-winter was acquired in Greenland on 12th January 1884.

Larus pacificus (Pacific Gull)
Endemic to the south coast of Australia and Tasmania. Monotypic. The single juvenile is devoid of collecting data.

Larus canus (Common Gull)
Palearctic and northwestern Nearctic. Four races. The nine skins represent a juvenile moulting to first-winter, first-winter, second-winter and adult plumages. With the exception of a specimen that is devoid of collecting data the others originate from localities within the British Isles.

Larus argentatus (Herring Gull)
Holarctic. At least five races. The seven specimens represent all plumage types. With the exception of a specimen without collecting data the others stem from British localities and were acquired during the following months; September, October, December, January, February and March.

Larus cachinnans (Yellow-legged Gull)
Western Palearctic (southern only). At least four races. The single juvenile, collected at Sarepta, Kalmytskaya, Russia on 2nd September 1884, has, in the past, caused identification problems. The specimen was originally labelled as *Larus ichthyaetus* (Great Black-headed Gull) but, at a later date this was questioned and a label stating that it is almost certainly *Larus cachinnans* added. Modern day knowledge has now made identification less of a problem.

Larus fuscus (Lesser Black-backed Gull)
Northwestern Palearctic. Three races. The five skins represent the following plumages; juvenile (1), sub-adult (1) and adult (3). The three

specimens with locality data originate from Bridlington, Yorkshire in October 1887 (adult); Norway in June 1888 (sub-adult) and near Southport, Lancashire on 2nd August 1938 (adult). The three adults belong to the race *graellsii* whilst the sub-adult, although labelled as belonging to the nominate race is probably of the race *intermedius*.

Larus marinus (Great Black-backed Gull)
Northern Atlantic. Monotypic. Seven specimens. The specimens with data (6) originate from Spurn, Yorkshire (three adults and a first-winter); the Shetland Islands, Scotland (adult) and Nolsoy hin, Faeroe Islands (sub-adult). The specimen without collecting data is in first-winter plumage.

Larus hyperboreus (Glaucous Gull)
Northern Holarctic. Three races. Eight specimens. The specimens with data (7) originate from Spurn, Yorkshire in December 1882 (first-winter), January 1916 (second-winter) and December of an unrecorded year (adult); Nolsoy hin, Faeroe Islands on 16th November 1946 (first-winter), 7th January 1947 (adult) and 19th January 1947 (first-winter). The specimen without locality data is in adult plumage and was collected on 11th October 1950.

Larus glaucoides (Iceland Gull)
Northern Nearctic (southern Baffin Island and Greenland only). Two races. The single adult, which belongs to the nominate race, was taken at Nolsoy, Faeroe Islands on 7th January 1947.

Larus pipixcan (Franklin's Gull)
Nearctic. Monotypic. The single adult, which exhibits full summer plumage, is devoid of collecting data.

Larus melanocephalus (Mediterranean Gull)
Western Palearctic (mainly Mediterranean and Black Sea). Monotypic. The two adults, both in full summer plumage, were taken at an untraced locality on an unrecorded date.

Larus ridibundus (Black-headed Gull)
Palearctic. Monotypic. The nine specimens represent the plumages of juvenile, first-winter and adult (both summer and winter). The specimens all originate from localities within the British Isles.

Larus genei (Slender-billed Gull)
Patchily distributed in the western and central Palearctic. Monotypic.
The single specimen, a downy chick, was collected in Turkey on 16th
June 1872.

Larus philadelphia (Bonaparte's Gull)
Northern Nearctic. Monotypic. Two specimens, both of which exhibit
full summer plumage. One specimen was collected at Victoria, British
Columbia, Canada on 6th May 1894 whilst the other is devoid of data.

Larus minutus (Little Gull)
Palearctic and, recently, Nearctic. Monotypic. The single specimen, a
juvenile, bears no more precise locality data than "southern Russia, 10th
August 1885".

Rhodostethia rosea (Ross's Gull)
Northeastern Siberia (occasionally Greenland). Monotypic. The single
adult was collected in Kamchatskaya Oblast, Russia in May or June 1907.

Rissa tridactyla (Kittiwake)
Northern Holarctic. Two races. The three specimens represent two adults
(both of which bear no collecting data) and a first-winter that was taken
at Scarborough, Yorkshire in January 1888.

Xema sabini (Sabine's Gull)
Northern Holarctic. Monotypic. The single juvenile was collected at
Bridlington, Yorkshire on 2nd December 1907.

STERNIDAE

Chlidonias leucoptera (White-winged Black Tern)
Central and eastern Palearctic. Monotypic. Four specimens, representing two adults and two immatures. The adults, which are both in full
summer plumage, are devoid of locality data but were collected in May
1876; the immatures were taken along the Volga River, Kalmytskaya,
Russia in August 1876 and Abu Labals, Egypt on 1st September 1917.

Chlidonias nigra (Black Tern)
Western and central Palearctic and Nearctic. Two races. The three skins,
all of which exhibit full summer plumage, originate from Hungary on
17th May 1883 and Italy on 25th April 1933 and during 1935.

Gelochelidon nilotica (Gull-billed Tern)
Almost cosmopolitan (save for Australasian). Six races. The three adults originate from Egypt, Africa on 19th November 1886; Entebbe, Uganda, Africa on 5th March 1901 and an untraced locality on 11th May 1874.

Sterna caspia (Caspian Tern)
Almost cosmopolitan (save for Neotropical). Monotypic. The single specimen, an adult in full breeding dress, originates from Tilcon Quarries, Catterick, Yorkshire on 1st July 1982. This skin is of interest as it is proof of an hitherto unpublished Yorkshire occurrence. The bird in question had been watched for several hours as it fished the area, but became entangled in discarded fishing line and was found dead, presumed drowned. A description to substantiate this record has never been submitted to the relevant authorities (British Birds Rarities Committee and Yorkshire Naturalists' Union Reports Committee) and, therefore, the record is unknown to the ornithological world. This will be only the fifteenth recorded occurrence of this species in Yorkshire and a full description is under preparation for the necessary authorities.

Sterna hirundo (Common Tern)
Holarctic. Three or four races. 14 specimens representing first-years and adults (in both summer and winter plumages). The specimens with data (9) originate from the Britsh Isles with the exception of an adult from Guernsey, Channel Islands on 29th June 1877.

Sterna paradisaea (Arctic Tern)
Northern Holarctic. Monotypic. 11 skins representing pullus, first-years and adults (in both summer and winter plumages). The specimens with data (8) originate from Spurn, Yorkshire (adult on 15th September 1882 and first-year on 2nd October 1900); Iceland (adult on 13th June 1874 and downy chick on 10th July 1885); Nolsoy hin, Faeroe Islands (adult on 30th June 1946 and two adults on 9th July 1946) and Spitsbergen (adult on an unrecorded date).

Sterna albifrons (Little Tern)
Cosmopolitan, though patchy in South America. Seven or eight races, although some American races are now considered a separate species (*Sterna antillarum*). The 12 specimens represent both juvenile and adult plumages. Of the 11 specimens with data, eight originate from the British Isles (Lancashire and Yorkshire). The other three (all adults) were collected at Pisa, Italy on 21st May 1937; Tangier, Morocco, Africa on an unrecorded date and at an untraced foreign locality on 3rd August 1926.

Sterna maximus (Royal Tern)
Southern Nearctic, northern Neotropical (patchily), Ethiopian (very patchily) and Palearctic (Mauritania only). Two races. The single specimen, an adult in non-breeding plumage, was taken near Port Royal, Jamaica, West Indies on 11th July 1891. The bird had been sexed as female upon dissection.

Sterna bengalensis (Lesser Crested Tern)
Southwest Pacific and Indian Oceans, extending as far west as the Red Sea and Mediterranean Sea (isolated colonies only). Two races. The two adult specimens originate from Morocco, Africa on 12th January 1888 and 16th January 1888.

Sterna sandvicensis (Sandwich Tern)
Western Palearctic, Nearctic and Neotropical (patchily). Three races, although some authors now regard *eurygnata* as a separate species. The single specimen, a first-winter, was taken in Albania on 25th February 1881.

ALCIDAE

Alle alle (Little Auk)
Northern Holarctic. Two races. The eight specimens, all of which are in winter plumage, were acquired from British and Irish coastal waters with the exception of a single from an indecipherable foreign locality (collected on 14th March 1884) and one from Helmsley, Yorkshire on 11th November 1953. Inland occurrences of this species, as indicated by the Yorkshire bird, are unusual; strong winds from the northern quarter occasionally cause inland 'wrecks' of this maritime species.

Alca torda (Razorbill)
North Atlantic. Two or three races. The five specimens represent a half-grown juvenile and four full-grown birds. The juvenile was taken at sea between Girvan and Ailsa Craig, Scotland on 10th August 1881 whilst the others were acquired on the Yorkshire coast (2); Lerwick, Shetland, Scotland (1) and the Faeroe Islands (1).

Uria aalge (Guillemot)
Holarctic. Up to seven races described. The 25 skins represent the plumages of juvenile and birds in summer and winter dress (including four belonging to the bridled morph). With the exception of eight specimens from Nolsoy hin, Faeroe Islands, the others originate from British

coastal waters (Cumberland (1), Shetland (1) and Yorkshire (15)). The two races, *aalge* and *albionis*, are represented in the sample.

Cepphus grylle (Black Guillemot)
Northern Holarctic. Up to six races described. The three specimens represent first-year, second-year and adult plumages. Only one specimen bears locality data, the first-year, which was collected at Ross, Dumfries and Galloway, Scotland in December 1947.

Fratercula arctica (Puffin)
North Atlantic. Three races. 11 specimens. Of the nine specimens with locality data, three originate from the Yorkshire coast (January 1881, autumn 1888 and February 1953) whilst the others were collected at Nolsoy hin, Faeroe Islands in July 1946 (5) and January 1947 (1).

COLUMBIFORMES

PTEROCLIDIDAE

Syrrhaptes paradoxus (Pallas's Sandgrouse)
Central Palearctic. Monotypic. The single specimen, a male, was taken at Easington, Yorkshire on 26th May 1888.

Pterocles alchata (Pin-tailed Sandgrouse)
Western and central Palearctic. Two races. The two specimens, a juvenile and a female, were taken at Arles, France on 9th September 1885 (female) and 10th September 1885 (juvenile).

Pterocles orientalis (Black-bellied Sandgrouse)
Southern Palearctic. Two or three races. The single specimen, a female, bears no more precise data than "south Russia, 6th August 1884".

COLUMBIDAE

Columba livia (Rock Dove)
Originally southern Palearctic, Oriental and northern Ethiopian, but now almost cosmopolitan due to feral populations. Up to 14 races described. Of the ten specimens only five show a lack of Man's interference ie. birds which appear to be pure bred. These originate from Fuertaventura, Canary Isles in March 1889; Isle of May, Scotland on 5th March 1948; Faeroe Islands on 10th March 1946 and 26th January 1948 and an untraced locality in the February of an unrecorded year. The bird

from the Canary Isles exhibits a grey rump, a phenomenon that can be shown by birds from these islands.

Columba oenas (Stock Dove)
Western Palearctic. Two or three races. Four specimens. With the exception of a skin that is devoid of collecting data the other three originate from English localities.

Columba palumbus (Woodpigeon)
Western Palearctic with isolated populations in central Palearctic. Five or six races. The five skins, representing the plumages of pullus (1), first-year (1) and adult (3), were collected at English localities. A specimen from Hertfordshire in April 1905 is showing albinistic tendencies, having a white rump and tail (save for the outer feather on each side which is normal) and white flecking on the mantle and wing coverts.

Ectopistes migratorius (Passenger Pigeon)
Formerly the whole of North America, but now extinct. The last known bird, a female named Martha, died in Cincinnati Zoo on 1st September 1914; no wild birds had been reported for 12 years prior to this event. The three specimens represent an adult male, an adult female and a juvenile male. Unfortunately none of the birds has collecting data although the adult male has a recently attached label which reads *"One of several specimens from the Knowsley menagerie of Low Derby. Purchased live by J.J.Audubon at New York in 1830 and lived at Knowsley for several years"*.

Streptopelia turtur (Turtle Dove)
Western and central Palearctic and, marginally, Ethiopian. Four races. Three specimens. With the exception of a specimen that is devoid of data the others were taken at Helmsley, Yorkshire on 17th June 1946 and 29th August 1948.

Streptopelia orientalis (Rufous Turtle Dove)
Central and eastern Palearctic and northern Oriental. Six races. The Yorkshire Museum has long been quoted as possessing the skin of the first British example of this species. The bird in question came from Scarborough, Yorkshire on 23rd October 1889 and was identified as this species by Henry Seebohm who also ascribed it to a bird in first plumage and probably of wild origin. The specimen was presented to the museum by James Backhouse. Although it is known that the museum was the custodian of this specimen it is no longer to be found on the premises

and has, apparently, not been seen by an ornithologist for more than 20 years!

Streptopelia chinensis (Spotted Dove)
Oriental. At least seven races. The single specimen was acquired at Foochow, China on 18th January 1885.

PSITTACIFORMES

PSITTACIDAE

Neophema bourkii (Bourke's Parrot)
Interior of central and south Australia. Monotypic. The single male specimen is without data and may well have originated from captive stock.

CUCULIFORMES

CUCULIDAE

Cuculus canorus (Cuckoo)
Palearctic and northern Oriental. Up to nine races described. Four specimens; a recently fledged juvenile is without collecting data, a juvenile and an adult were collected at Colchester, Essex during 1863 and an adult was taken at York, Yorkshire on 20th June 1918. Both adults are of the grey morph.

Coccyzus americanus (Yellow-billed Cuckoo)
Nearctic and northern Neotropical. Two races. The two specimens are devoid of collecting data.

STRIGIFORMES

TYTONIDAE

Tyto alba (Barn Owl)
Cosmopolitan (save for central and eastern Palearctic). At least 36 races. 13 skins representing a juvenile and both sexes. With the exception of specimens from Morocco, Africa on 12th October 1886 and an untraced foreign locality on 9th November 1886, the others with collecting data (9) originate from British localities;· Isle of Wight (female); Northamptonshire (female); Nottinghamshire (male); Cheshire (juvenile) and Yorkshire (male and four females).

STRIGIDAE

Otus scops (Scops Owl)
Palearctic and, marginally, Oriental. Up to 21 races described. The two specimens originate from southern France on 12th March 1900 and an untraced locality on 26th May 1889.

Bubo bubo (Eagle Owl)
Palearctic, Oriental and, marginally, northern Ethiopian. At least 21 races. The two specimens were collected at Nordmore, Norway on 6th March 1925 and 20th September 1925.

Surnia ulula (Hawk Owl)
Northern Holarctic. Three races. The single specimen is devoid of collecting data.

Athene noctua (Little Owl)
Palearctic and northeastern Ethiopian. At least 16 races. 12 specimens. Of the ten specimens with data, eight were taken at British localities. The other two were acquired in Morocco, Africa on 1st June 1884 and Jordan on 26th June 1895. Both these foreign specimens are very pale; the one from Morocco probably belonging to the race *glaux*.

Strix aluco (Tawny Owl)
Palearctic and northern Oriental. 12 or 13 races. The three specimens were acquired at localities within Yorkshire. One shows a tawny-chestnut ground colour whilst the other two are a deep rufous.

Asio otus (Long-eared Owl)
Holarctic and Ethiopian (patchily). Up to six races described. The three specimens originate from Yorkshire localities; Aberford in 1880 and Spurn on 20th November 1886 and 30th July 1887.

Asio flammeus (Short-eared Owl)
Holarctic and Neotropical. Nine or ten races. Six specimens. With the exception of two specimens from an untraced locality on unrecorded dates, the others were collected in Colchester, Essex on 29th October 1878; Spurn, Yorkshire on 20th October 1900 and in December of an unrecorded year and Edeyrn, Caernarvonshire, Wales on 4th February 1927.

44

Asio capensis (Marsh Owl)
Ethiopian and southwestern Palearctic (Morocco only). Three races. The two specimens were both collected in Morocco, Africa (one on 25th June 1887, the other on an unrecorded date).

Aegolius funereus (Tengmalm's Owl)
Northern Holarctic. Seven or eight races. Four specimens. With the exception of a specimen which bears no data the other three originate from Surrendal, Norway and were taken on 13th March 1925, 6th January 1926 and 13th October 1926.

CAPRIMULGIFORMES

CAPRIMULGIDAE

Caprimulgus europaeus (Nightjar)
Western and central Palearctic. Up to ten races described. The four skins represent juvenile and female plumages. The three specimens with data (a juvenile and two females) were taken at localities in Great Britain and presumably belong to the nominate race.

APODIFORMES

APODIDAE

Apus melba (Alpine Swift)
Southern Palearctic, Oriental and Ethiopian. At least ten races. The single specimen was taken in the south of France on 2nd June 1886.

Apus pallidus (Pallid Swift)
Southwestern Palearctic, northeastern Ethiopian and northwestern Oriental. Three races. The single specimen bears no locality data but was collected on 20th May 1884.

Apus apus (Swift)
Palearctic. Two races. With the exception of a specimen which bears no data, the other six were taken at British localities.

Apus unicolor (Plain Swift)
Endemic to Madeira and the Canary Isles. Monotypic. The single specimen bears no collecting data.

Apus affinis (Little or House Swift)
Ethiopian, Oriental and extreme southwestern and eastern Palearctic.
At least 12 races. The three specimens originate from Syria (1) on 20th
May 1883 and northern India (2) on unrecorded dates.

CORACIIFORMES

ALCEDINIDAE

Alcedo atthis (Kingfisher)
Palearctic, Oriental and northern Australasian. At least eight races. Of
the seven specimens only four carry collecting data; Spurn, Yorkshire on
24th August 1885 and early autumn 1878; Normandy, France on 7th
March 1872 and the Yangtse River, China in 1878.

Halcyon pileata (Black-headed Kingfisher)
Oriental (mainly coastal). Monotypic. The single specimen, sexed as
male upon dissection, was taken at Foochow, China on 19th February
1887.

MEROPIDAE

Merops orientalis (Little Green Bee-eater)
Oriental, northern Ethiopian and, marginally, southern Palearctic. At
least seven races. The single specimen was collected in the Punjab (India
or Pakistan) on an unrecorded date.

Merops superciliosus (Blue-cheeked Bee-eater)
Oriental, Ethiopian (patchily) and southern Palearctic (patchily). Three
or four races. Of the three skins, which are all in adult plumage, two
originate from untraced localities (collected 11th May 1877 and an
unrecorded date) whilst the other bears no locality data but was acquired
on the 28th March of an unrecorded year. The labels attached to the specimen lacking a collecting date states that it had been collected at "Ghaba
Nambs, Spain". This species has not previously been recorded in Spain
and, on the surface, this would appear to be the first record. However,
the ornithological authorities in Spain (Sociedad Espanola de
Ornitologia) indicate that as there is no locality of that name in Spain, a
mistake has occurred. As the locality and the country are on separate
labels (in different handwriting) it is obvious that an error has taken
place and I can do no more than reflect this sentiment. Investigations
have revealed that there are a number of localities in the Middle East

which begin with the name 'Ghaba' (an area from which the specimen is far more likely to originate) and it is suggested that "Spain" on the label is an unfortunate and misleading error.

Merops apiaster (Bee-eater)
Western Palearctic, northwestern Oriental and Ethiopian (south Africa only). Monotypic. The three specimens, which represent a juvenile (sexed as male upon dissection) and two adults (sexed as male and female upon dissection) were collected in Turkey (purported female) in April 1875, Cyprus (juvenile) on 11th August 1883 and southern Spain (purported male) on 18th May 1884.

CORACIIDAE

Coracias garrulus (Roller)
Western and central Palearctic. Two races. The five adult specimens originate from Bulgaria in June 1866 (2); Sarepta, Kalmytskaya, Russia in May 1876 (2), whilst the other contains no more precise collecting data than "southern Russia".

UPUPIDAE

Upupa epops (Hoopoe)
Palearctic, Oriental and Ethiopian. Ten races. The three specimens were acquired in Italy on 10th April 1877 (2) and at an untraced locality in 1877.

PICIFORMES

PICIDAE

Jynx torquilla (Wryneck)
Palearctic. Up to ten races described. 11 specimens, the seven with data originate from Nice, France (2), Corsica (2) and Great Britain (3).

Dendrocopus minor (Lesser Spotted Woodpecker)
Palearctic. Up to 19 races described. Five specimens; a juvenile from Great Britain; a male and female from southern France and a male and female from an untraced locality.

Dendrocopus leucotos (White-backed Woodpecker)
Palearctic. At least 14 races. Four specimens representing both sexes. With the exception of a male and female from Sweden (male on 7th

January 1861 and female on an unrecorded date) the others, two females, are from untraced localities. The Swedish male was figured by C.R.Bree in his book *A History of the Birds of Europe, not observed in the British Isles* (1875).

Dendrocopus medius (Middle Spotted Woodpecker)
Western Palearctic. Up to seven races. Three specimens. With the exception of a specimen from Turkey on 23rd February 1882, the others originate from untraced localities on 21st August 1882 and 26th June 1883. This latter mentioned bird had, upon dissection, been identified as a female.

Dendrocopus syriacus (Syrian Woodpecker)
Southern central Palearctic. Three or four races. The single juvenile specimen bears no collecting data.

Dendrocopus major (Great Spotted Woodpecker)
Palearctic and, marginally, Oriental. Up to 31 races described. The 12 specimens represent both sexes and juvenile plumages, the countries of origin being Great Britain (two juveniles); Switzerland (male); Sweden (male); Latvia (male and female); Prussia (juvenile) and Morocco, Africa (male). The other specimens have untraced localities (male and female) or bear no data (two females). The specimen from Morocco, collected on 10th April 1894, belongs to the race *mauritanus*.

Picoides tridactylus (Three-toed Woodpecker)
Holarctic. Up to 11 races described. Of the five specimens (three males and two females) only two carry locality data; a male from Sweden in November 1860 and a female which bears no more precise collecting data than "northern Russia, 7th May 1899". The Swedish bird was figured by C.R.Bree in his book *A History of the Birds of Europe, not observed in the British Isles* (1875).

Dryocopus pileatus (Pileated Woodpecker)
Nearctic. Four races. The single male specimen is simply labelled "Switzerland, 1876". As this species is extremely unlikely to occur on this side of the Atlantic the label is either erroneous or there is an untraced locality of that name within the species' range (see *Porzona carolina* above).

48

Picus awokera (Japanese Green Woodpecker)
Endemic to Japan. Three races. The two specimens, representing both
sexes, were taken in Japan on unrecorded dates.

Picus canus (Grey-headed Woodpecker)
Palearctic and eastern Oriental. Up to 16 races described. Four specimens (three males and a female). The female, collected on 7th October
1884, is from an untraced locality. Of the males, one was taken in Sweden
on 1860; one was acquired in Hannover, Germany in July 1882 and the
other has a label indicating that it was collected at Taschkent, Ubekskaya,
Russia in 1883. The Swedish bird was figured by C.R.Bree in his book *A
History of the Birds of Europe, not observed in the British Isles* (1875). The
Ubekskaya (formerly Turkestan) skin has a note attached by D.T. Lees-
Smith which indicates that the locality details are erroneous as the species
does not occur in Ubekskaya. He goes on to add that the specimen was
collected along the route of the Trans-Siberian Railway (the species is
present in southern Siberia) and somehow became mixed up with birds
collected in the general area of Ubekskaya. I can do no more than reflect
this sentiment.

Picus viridis (Green Woodpecker)
Western Palearctic. Up to 11 races described. Of the six specimens, which
represent both sexes, only two bear locality data; a male from southern
France on 1st April 1881 and a female from Osbaldwick, Yorkshire on
30th November 1918.

PASSERIFORMES

ALAUDIDAE

Melanocorypha calandra (Calandra Lark)
Southern Palearctic. Three races. The four specimens were taken in
France in January 1886 (3) and Morocco, Africa on an unrecorded date
(1). The French birds were purchased in the markets at Montpellier (2)
and Perpignan (1).

Melanocorypha leucoptera (White-winged Lark)
Central Palearctic. Monotypic. The three specimens, two males and a
female, bear no more precise locality data than "southern Russia". One
of the males was acquired on 21st December 1882, the female on 2nd
December 1887 whilst the other male is devoid of a collecting date.

Melanocorypha yeltoniensis (Black Lark)
Central Palearctic. Monotypic. The breeding plumaged male and a female were collected in southern Russia; the male at Sarepta, Kalmytskaya on 5th May 1894 and the female from an untraced locality on 12th December 1890.

Calandrella brachydactyla (Short-toed Lark)
Western and central Palearctic. Up to 20 races described. Ten specimens. With the exception of a specimen that is devoid of locality data the others originate from Spain (2); France (2); Italy (1); Malta (1); Switzerland (1); Turkey (1) and Britain (1). The British record is of interest as the species is but a rare visitor to the British Isles. Unfortunately the bird in question is simply labelled "British" and the very broad geographical distribution dictates that it cannot be added to any particular county list. The specimen from Switzerland, collected at an unrecorded locality in 1876, is also of interest as the species is but an accidental visitor to that country.

Calandrella rufescens (Lesser Short-toed Lark)
Southern Palearctic. At least 16 races. The single specimen was collected at Sarepta, Kalmytskaya, Russia on 20th May 1887.

Galerida cristata (Crested Lark)
Palearctic, northern Ethiopian and, marginally, Oriental. Up to 28 races described. The nine specimens originate from France (2); Switzerland (2); Germany (3); Turkey (1) and Tunisia, Africa (1).

Lullula arborea (Woodlark)
Western Palearctic. Two races. Seven specimens. With the exception of a specimen that is devoid of locality data the others originate from Cornwall (1); Yorkshire (1); France (2) and Corsica (2). The Yorkshire bird, from Easington on 1st March 1888, is of interest as the record was not published in Nelson's *The Birds of Yorkshire* (1907) or Chislett's *Yorkshire Birds* (1952). The specimen adds an hitherto unrecorded occurrence to the Yorkshire list.

Alauda arvensis (Skylark)
Palearctic (introduced to southwestern Australia and New Zealand). At least 16 races. The 34 skins represent the plumages of juvenile and full-grown birds. With the exception of three specimens that are lacking locality data the others originate from the British Isles (27); Ireland (1); France (2) and Kiang-su Province, China (1).

50

Eremophila alpestris (Shore Lark)
Holarctic and Neotropical (patchily). Up to 41 races described. 27 specimens. With the exception of two specimens that are devoid of locality
data the others were acquired in England (13); Norway (1); Ust Tsilma,
Russia (2); southern Russia (1); the Karakorums, Kashmir, India (6) and
two from untraced foreign localities. A specimen from Lilford (most
probably from Lord Lilford's menagerie at Lilford Hall, near Oundle,
Northampton and not the Lilford in Greater Manchester) shows bright
yellow head markings and may belong to the race *flava*. The wide geographical provenance of the specimens ensures that a good number of
races are available.

HIRUNDINIDAE

Riparia riparia (Sand Martin)
Holarctic and Oriental. At least five races. The two adults and a first-
winter were collected at British localities.

Ptyonoprogne rupestris (Crag Martin)
Southern Palearctic and northeastern Ethiopian (Saudi Arabia only).
Monotypic. The three specimens originate from Pinerolo, Italy (2) on an
unrecorded date and Corsica (1) on 27th December 1890.

Hirundo rustica (Swallow)
Holarctic and northern Oriental. Eight races. Two of the six specimens
belong to one of the races that exhibit rufous underparts; one was taken
in Egypt on 1st November 1891 whilst the other bears no collecting data.
The others (save for one without collecting data) originate from British
localities.

Hirundo daurica (Red-rumped Swallow)
Southern Palearctic, Oriental and Ethiopian (patchily). 11 or 12 races.
The two specimens originate from Turkey (a juvenile collected on 2nd
July 1886) and Malta (an adult collected on 8th April of an unrecorded
year).

Delichon urbica (House Martin)
Palearctic and northern Oriental. Five races. Seven specimens, those with
data (6) originate from British localities. Two specimens from Arlington,
Yorkshire on 13th May 1886 were killed by the inclemency of the weather
(see the *Naturalist* 1886: 182). One individual, collected at Heath End,
Tadley, Hampshire on 10th July 1936 shows completely leucistic
plumage.

MOTACILLIDAE

Motacilla flava (Yellow Wagtail)
Palearctic and northwest Nearctic. 18 races. The 48 specimens originate from as far afield as Europe and India. The more typical males represent the subspecies *flava, flavissima, thunbergi, feldegg and cinereocapilla*.

Motacilla citreola (Citrine Wagtail)
Central Palearctic. Three races. The four specimens, which are all males, originate from Ubekskaya, Russia on 23rd March 1930; Lahore, Pakistan (collecting date unknown); an unrecorded locality in March 1870 and an untraced locality on 5th June 1875.

Motacilla cinerea (Grey Wagtail)
Palearctic. Five or six races. Eight of the 15 specimens originate from localities in the British Isles. The others, with the exception of one lacking locality data, were collected in Turkey (male on 16th March 1867 and 20th January 1879); France (male on 30th May 1883); Tenerife, Canary Isles (summer plumaged male in 1887); China (29th November 1886) and Japan on an unrecorded date.

Motacilla alba (Pied Wagtail)
Palearctic, northern Oriental and extreme northwestern Nearctic. Up to 11 races described. 34 skins, most of which represent the race *yarrellii*. The specimens which belong to this race all originate from localities in the British Isles. Specimens belonging to the nominate race were acquired in Germany (4); Norway (1) and the Faeroe Islands (1). A specimen that is attributable to the race *alboides* was taken in Kashmir, India on 31st May 1904. A female of the race *yarrellii* from Barmouth, Merionethshire, Wales on 27th September 1899 had been shot with a catapult!

Motacilla maderaspatensis (Large Pied Wagtail)
Restricted to India and Pakistan. Monotypic. The single specimen, collected in India, bears no other data (although its sex is stated to be female).

Motacilla flaviventris (Madagascar Wagtail)
Endemic to Madagascar. Monotypic. The two specimens were collected on Madagascar; a bird sexed as male upon dissection in June 1925, the other, sexed as female upon dissection, in May 1928.

Anthus richardi (Richard's Pipit)
Central and eastern Palearctic and northeastern Oriental. At least six races. The single specimen was collected in Sibirskaya, Russia on 15th July of an unrecorded year.

Anthus campestris (Tawny Pipit)
Western and central Palearctic. Three races. Four specimens. The three specimens with data originate from Turkey on 14th April 1879; Morocco, Africa on 17th May (no date appended) and Malta on 14th May (no date appended). The specimen without locality data was collected on 24th April 1878.

Anthus pratensis (Meadow Pipit)
Western Palearctic. Two races. 16 specimens. The specimens with locality data (15) originate from the British Isles with the exception of singles from Corsica (27th December 1890); Spain (10th January 1898); Faeroe Islands (19th May 1946); an untraced locality (29th October 1898) and two from Iceland (13th and 19th September 1884).

Anthus trivialis (Tree Pipit)
Western and central Palearctic. Three races. Four of the seven specimens were collected at British localities. The others originate from Italy (10th September 1937); an untraced locality (6th May 1889) and one devoid of locality data (22nd August 1897).

Anthus hodgsoni (Olive-backed Pipit)
Central and eastern Palearctic. Two races. The two specimens were acquired in Tokyo, Japan on an unrecorded date and in the Himalayas (no country or date appended).

Anthus roseatus (Hodgson's Pipit)
Northern Oriental and extreme southeastern Palearctic. Monotypic. The two specimens were collected in Kashmir, India; one on 27th January 1907 the other on 18th September 1908.

Anthus cervinus (Red-throated Pipit)
Northern Palearctic. Monotypic. The four specimens originate from Turkey on 18th April 1879; Egypt, Africa on an unrecorded date; Siberia on an unrecorded date and an untraced locality on 17th May (no year appended).

53

Anthus petrosus (Rock Pipit)
Palearctic. At least two, probably more races. 34 specimens. With the exception of specimens from southern France (1) and the Faeroe Islands (4), the others were all acquired at localities in the British Isles.

Anthus spinoletta (Water Pipit)
Palearctic. Three races. 11 specimens. With the exception of a specimen devoid of locality data, the others were taken in the Pyrenees, France (2); Spain (2); Switzerland (2); Cyprus (1); Turkey (1) and Great Britain (2). The British specimens were collected at Tetney, Lincolnshire on 5th April 1895 and Treeth Mawr, Caernarvonshire, Wales on 5th April 1897. As most of the specimens were acquired in spring they are exhibiting partial or full summer plumage.

Anthus rubescens (Buff-bellied Pipit)
Nearctic and extreme eastern Palearctic. Two races. The single specimen was collected in Greenland on an unrecorded date.

Anthus berthelotii (Berthelot's Pipit)
Endemic to Madeira and the Canary Isles. Two races. The single specimen was collected on Tenerife, Canary Isles on 9th February 1887.

PYCNONOTIDAE

Pycnonotus xanthopygos (Yellow-vented Bulbul)
Eastern Mediterranian and Saudi Arabia. Monotypic. The single specimen, sexed as male upon dissection, was acquired in Beirut, Lebanon on 24th April 1891.

Pycnonotus barbatus (Common Bulbul)
Southwestern Palearctic (north Africa only) and northern Ethiopian. Up to 18 races described. Two specimens. One was taken in Tangier, Morocco, Africa and the other has indecipherable locality data. Neither specimen bears a collecting date.

LANIIDAE

Tchagra senegala (Black-headed Bush Shrike)
Ethiopian and Palearctic (confined to Morocco, northern Algeria and northern Tunisia). Up to 15 races described. The five specimens were collected in Africa; Morocco (3), Ethiopia (1) and southern Malawi (1).

Lanius tigrinus (Tiger Shrike)
Extreme eastern Palearctic and northeastern Oriental. Monotypic. The four skins, which represent a male, two females and a juvenile, originate from China (female in May 1888), Korea (male and female in May 1888) and Indonesia (juvenile on 20th December 1887).

Lanius bucephalus (Bull-headed Shrike)
Extreme eastern Palearctic and northeastern Oriental. Two races. Four of the five specimens originate from Yokohama, Japan on an unrecorded date whilst the other is from Seoul, Korea on 23rd March 1889; both sexes are represented in the sample.

Lanius cristatus (Brown Shrike)
Extreme southeastern Palearctic and northern Oriental. Four races. The 26 specimens originate from China, Korea, Indonesia, Taiwan, Japan and Siberia. Skins previously examined by D.T. Lees-Smith have been ascribed to the races *cristatus* (9), *lucionensis* (10) and *supersiliosus* (6).

Lanius collurio (Red-backed Shrike)
Western and central Palearctic. Three races. 23 skins, encompassing all age groups and both sexes. Several of the specimens were collected at British breeding localities (between 1883 and 1925) with birds taken on their continental breeding grounds being represented by specimens from: Sweden, Czechoslovakia, Hungary, Turkey, East Germany, France, Italy and Spain. Although now nearly extinct as a British breeding bird, a pullus taken at Neath, Glamorganshire, Wales on 15th July 1883, is an indication of the bird's former breeding range. Three birds, taken during November 1881, originate from the species' wintering grounds in South Africa.

Lanius isabellinus (Isabelline Shrike)
Southern central Palearctic. Four races. Five specimens. With the exception of a specimen from northern India on 7th January 1883 the others originate from the Altai region of Russia (collected on 14th May 1883, 2nd May 1886, 11th May 1886 and 3rd April 1889). These skins had previously been examined by D.T. Lees-Smith and ascribed to the nominate race (Indian bird and a Russian bird) and *phoenicuroides* (three Russian birds).

Lanius schach (Rufous-backed or Long-tailed Shrike)
Southeastern Palearctic and Oriental. Up to 14 races described. The 15 specimens were acquired in Burma (1); China (8); India (4); Indonesia

(1) and Ubekskaya, Russia (1). These skins had previously been examined by D.T. Lees-Smith and ascribed to the races *caniceps* (1), *erythronotus* (5) and *schach* (9). A specimen from southern China on 25th April 1860, although ascribed to the nominate race by D.T. Lees-Smith, also has a note attached by him indicating that the *"Invasion of hind crown and shoulders by black from crown and 'mask' show integration towards L. schach tricolor; Hodgson 1837"*.

Lanius minor (Lesser Grey Shrike)
Western and central Palearctic. Monotypic. The six specimens originate from Austria (juvenile), southern France (juvenile), Hungary (adult), Sarepta, Kalmytskaya, Russia (two juveniles) and the Lower Volga, Russia (adult). One of the specimens from Sarepta, (no date attached but probably late 19th century) is labelled "Lanius major (Pall) *juv.? The only thing one can be sure of is that this is a young Great Grey!!"*. This was, one must remember, long before wing formula and primary projection had been recognized as identification features.

Lanius ludovicianus (Loggerhead Shrike)
Southern Nearctic. 11 races. The single specimen, which had been sexed as female upon dissection, was taken at Puebla, Mexico, Central America on an unrecorded date.

Lanius excubitor (Great Grey Shrike)
Palearctic, Oriental and northern Ethiopian. Up to 19 races described. A well represented species with 47 specimens encompassing races from Europe, north Africa and Russia. These skins had previously been examined by R.Wagstaffe and E.W.Taylor and ascribed to the races *excubitor* (10); *homeyeri* (8); *meridionalis* (6); *koenigi* (8); *algeriensis* (2); *elegans* (4) and *dodsoni* (1).

Lanius sphenocercus (Chinese Great Grey Shrike)
Northeastern Oriental. Two races. The two specimens were collected from untraced localities on 21st December 1885 and 20th September 1888.

Lanius senator (Woodchat)
Southwestern Palearctic. Four races. The 20 specimens originate from the following countries: France (one collected in April and two in May), Italy (April, July and undated), Malta (two in April and one undated), Spain (two in May), West Germany (June), Israel (March), Morocco, Africa (March, April, May and June), Tunisia, Africa (July) and the

Gambia, Africa (two undated). One of the birds collected in the Gambia is still largely in juvenile plumage. The species does not breed south of the Sahara, the nearest breeding grounds being in southern Morocco, so the bird was obviously on passage. The moult cycle of this species is complicated, but it has been suggested that individuals which migrate whilst undertaking their post-juvenile moult are probably late-hatched birds.

Lanius nubicus (Masked Shrike)
Western Palearctic (confined to the eastern Mediterranean as far east as western Turkey). Monotypic. The six specimens originate from Turkey (juvenile on 3rd July 1879, male on 12th April 1880 and a juvenile on 19th June 1883) and Egypt, Africa (two males and a female on unrecorded dates).

BOMBYCILLIDAE

Bombycilla garrulus (Waxwing)
Holarctic. Three races. The 19 skins represent both sexes in first-year and adult plumages. The only specimens with data were taken at Shanghai, China in December 1894; Sweden in December 1881 and December 1885; Latvia in December 1924 (2) and Blakeney Point, Norfolk in March 1928. The county of origin of this last mentioned bird is erroneously given as "Suffolk".

Bombycilla japonica (Japanese Waxwing)
Extreme southeastern Palearctic and northeastern Oriental. Monotypic. The single specimen was acquired at Yokohama, Japan on 8th April 1882.

CINCLIDAE

Cinclus cinclus (Dipper)
Palearctic. Up to 11 races described. The 21 specimens represent juvenile, first-year and adult plumages. With the exception of an adult that is devoid of locality data, the others originate from the British Isles (juvenile and adults); Pyrenees, France (juveniles and adult); Norway (first-year and adults); Sweden (juvenile and adults) and an unrecorded country in Lapland (juvenile). The wide geographical provenance of the specimens ensures that several races are available.

TROGLODYTIDAE

Troglodytes troglodytes (Wren)
Holarctic and, marginally, Oriental. Up to 37 races described. 18 specimens. The specimens with locality data (13) originate from the British Isles (9); Corsica (1); the Faeroe Islands (2) and an untraced locality (1).

PRUNELLIDAE

Prunella collaris (Alpine Accentor)
Southern Palearctic. Ten races. Of the five specimens only three carry collecting data; Andorra, Spain on 25th May 1889 and Switzerland on 20th February 1895 and on an unrecorded date.

Prunella modularis (Dunnock)
Western Palearctic. Eight races. The seven specimens originate from British localities with the exception of an adult from the Pyrenees, France on 1st June 1896.

TURDIDAE

Cercotrichas galactotes (Rufous Bushchat)
Southern Palearctic, northern Ethiopian and, marginally, Oriental. Five or six races. The seven specimens originate from Morocco, Africa on 12th June 1890; Egypt, Africa on 14th and 15th April 1863; Turkey in May 1875, 18th May 1878 and 5th May 1894 and Gibraltar on an unrecorded date.

Erithacus rubecula (Robin)
Western Palearctic. Eight races. 12 specimens. With the exception of specimens from Austria in November 1885 and on 3rd December 1898 and Morocco, Africa on 18th April 1899 the others originate from the British Isles. An adult that had been ringed near Northwich, Cheshire on 18th January 1946 was accidentally caught in a mouse trap at the same locality on 9th December 1947 (the specimen still wears its British Trust for Ornithology ring, number LK746).

Luscinia luscinia (Thrush Nightingale)
Central Palearctic. Monotypic. Three specimens; two adults and a pullus. The two adults were acquired at Sarepta, Kalmytskaya, Russia on 9th July 1884 and Budapest, Hungary on 3rd September 1878. The pullus, which was collected in Sweden on 6th July 1883, is nearly fully grown and exhibits the wing formula indicative of the species.

Luscinia megarhynchos (Nightingale)
Southwestern Palearctic. Three races. Five specimens. The specimens
with data (3) originate from Essex, England; Saxony, Germany and Nice,
France.

Luscinia calliope (Siberian Rubythroat)
Central and eastern Palearctic. Monotypic. The single specimen. a
female, was taken at Kamchatka, Russia on 2nd July 1886.

Luscinia svecica (Bluethroat)
Palearctic and extreme northwestern Nearctic. Seven races. The 21 specimens (15 males and six females) all originate from European countries
(although a male from an untraced locality is not necessarily from this
area). A male of the nominate race was taken in Norway on 2nd June
1882. Breeding plumaged males of the race *cyanecula* were collected in
Switzerland (4) and southern France (2) with another simply being
labelled as "Europe". This latter mentioned bird, along with a female
bearing the same locality data, were figured by C.R.Bree in his book *A
History of the Birds of Europe, not observed in the British Isles* (1875).

Tarsiger cyanurus (Red-flanked Bluetail)
Central and eastern Palearctic and, marginally, Oriental. Two or three
races. Four specimens; two adult males and two female-types. The males
were collected in China on 28th December 1886 and Japan in November
1922 whilst the female-types were taken in China on 28th April 1884 and
at an unrecorded locality during the winter of 1887/1888.

Irania guttularis (White-throated Robin)
Extreme southern central Palearctic and northeastern Oriental.
Monotypic. The single specimen, a first-winter, was taken at Smyrna
(now Izmir), Turkey on 26th April 1880.

Phoenicurus ochruros (Black Redstart)
Western and central Palearctic and, marginally, Oriental. Up to seven
races described. The 14 specimens represent adult male and female-type
plumages. With the exception of two that are devoid of locality data and
a single from an untraced locality, the others originate from Great Britain
(3); France (3); Germany (1); Europe (2) and northern India (2). The two
specimens from northern India, both of which are males, belong to one
of the eastern races that exhibit chestnut on the underparts.

Phoenicurus phoenicurus (Redstart)
Western and central Palearctic. Two or three races. The 21 skins represent male (in both summer and winter dress), female and juvenile
plumages. The specimens with data (17) originate from British localities
with the exception of the following (all males); Arkhangel'sk, Russia on
1st July 1872; Hungary in September 1879; Norway on 27th June 1888
and Spain on 1st June (no year appended).

Phoenicurus fuliginosus (Plumbeous Redstart)
Northeastern Oriental and extreme southeastern Palearctic. Two races.
Three specimens representing both sexes; the two males were taken in
China on 17th January 1886 and an unrecorded date whilst the female
bears no locality data but was acquired on 28th December 1886.

Phoenicurus auroreus (Daurian Redstart)
Southeastern Palearctic and northeastern Oriental. Two races. Two male
specimens. One was taken at an untraced locality during the winter of
1887/1888 whilst the other is devoid of data.

Phoenicurus moussieri (Moussier's Redstart)
Endemic to Morocco, Algeria and Tunisia. Monotypic. The single specimen, a male, was taken in Morocco, Africa on 14th October 1888.

Saxicola rubetra (Whinchat)
Western Palearctic. Monotypic. Of the 16 specimens, which represent
both sexes, 12 originate from British localities. The others were collected
in Turkey (a male on 12th April 1879); Yugoslavia (a male on 1st June
1883) and two males from untraced foreign localities.

Saxicola torquata (Stonechat)
Palearctic, Ethiopian and, marginally, Oriental. 25 races. The wide geographical provenance of the 31 specimens (between Ireland in the west
and Japan in the east) ensures that a good number of the described
Palearctic races are available. The specimens with data (29) originate
from Britain and Ireland (11), Corsica (2), India (14) and Japan (2) and
probably represent the races *rubicola, hibernans, maura* and *stejnegeri*.

Saxicola leucura (White-tailed Stonechat)
Restricted to northern India and Pakistan. Monotypic. The single specimen, a female, was collected in the Punjab, India on the 24th January
1880.

Saxicola caprata (Pied Stonechat)
Oriental. 16 races. The single specimen, a male, was collected in Pakistan on 19th January 1880.

Oenanthe isabellina (Isabelline Wheatear)
Southern Palearctic and northwestern Oriental. Monotypic. Six specimens. The five specimens with locality data originate from Turkey (17th and 25th March 1879); Egypt, Africa (13th February 1918), whilst the other two carry no more precise data than "east Africa, 4th March 1898" and, a bird showing distinct male characters, from "southern Russia, 11th May 1883".

Oenanthe oenanthe (Wheatear)
Palearctic and northern Nearctic. Up to seven races described. 37 specimens representing male, female and juvenile plumages. With the exception of a single specimen the others all carry collecting data, of which 23 originate from British localities. The others were taken in France (2); Germany (1); Cyprus (2); Turkey (2); Norway (1); Faeroe Islands (4) and Greenland (1). Two birds, a male from Powys, Wales in April 1901 and an unsexable bird from Looe, Cornwall on 5th October 1909, can be attributed to the race *leucorhoa*. The bird from Greenland, a juvenile without a collecting date, presumably belongs to this race but, because of the bird's immaturity, biometrics are of little use.

Oenanthe deserti (Desert Wheatear)
Southern and central Palearctic. Three or four races. The four specimens represent male, female and juvenile plumages. The females are devoid of locality details but the juvenile was collected in Kashmir, India on 28th August (no date appended) and the male in Rajasthan, India on 4th February 1878.

Oenanthe hispanica (Black-eared Wheatear)
Southern Palearctic. Two races. The six specimens, all males, were acquired from the following areas; Turkey (3rd April 1882, 29th January 1884, 29th April 1893 and 6th May 1895), Morocco, Africa (10th March 1884) and Italy (31st March 1910).

Oenanthe finschii (Finsch's Wheatear)
Southern central Palearctic. Two races. The two specimens, both females, were acquired in Asia Minor on 4th March 1879 and 19th March of the same year.

Oenanthe picata (Eastern Pied Wheatear)
Extreme southern central Palearctic and western Oriental. Monotypic, but occurring in two colour morphs (polymorphic). Two specimens. A first-year male of the *opistholeuca* morph was collected at Sultanpur, India on 18th January 1879 and a first-year male of the *picata* morph was acquired at an untraced locality on 29th November 1877.

Oenanthe lugens (Mourning Wheatear)
Western Palearctic (southern only), northeastern Ethiopian and north-western Oriental. Up to eight races described. The two specimens originate from Malawi, Africa on 20th January 1894 and Egypt, Africa in 1895.

Oenanthe pleschanka (Pied Wheatear)
Central Palearctic. Two races. The single specimen, a male, was collected at Lahore, Pakistan on an unrecorded date.

Oenanthe leucopyga (White-crowned Black Wheatear)
Western Palearctic (north Africa only) and northern Ethiopian. Two or three races. The single specimen, an immature, bears no more precise data than "southern France, 21st April 1884". The record is of great significance, however, as it constitutes the first recorded occurrence of this species in France. The skin, identified as male upon dissection, had originally been labelled as *Saxicola leucura* (a synonym of *Oenanthe leucura* Black Wheatear). As the bird in question is in immature plumage, and therefore sporting a black cap, this mistake is easy to understand. The main identification feature between these two species lies in the tail pattern, and when the initial identification was undertaken this characteristic was obviously missed. Full details of the bird have been forwarded to the French Rarities Committee (Comite D'Homologation National) and their deliberations regarding this most interesting find are still awaited.

Oenanthe leucura (Black Wheatear)
Southwestern Palearctic. Two races. The single female was purchased at Perpignan market, France on 5th January 1886.

Oenanthe moesta (Red-rumped Wheatear)
Southwestern Palearctic (confined to north Africa). Two races. The two specimens, a male and female, were acquired in Tunisia, Africa; the female in 1894 and the male on an unrecorded date.

Monticola saxatilis (Rock Thrush)
Southern Palearctic. Monotypic, although some authors recognize two races. The five specimens originate from southern France (first-year on 29th September 1884 and a male on 29th May 1891); two untraced localities (a male on 7th May 1877 and an unsexable bird in 1879) and a male that is devoid of data.

Monticola solitarius (Blue Rock Thrush)
Southern Palearctic. Five races. Eight specimens (four males and four females). Five of the specimens were collected in Europe; France (two females) and Cyprus (a male and two females). A male from Foochow, China on 11th January 1887 belongs to the race *philippensis* whilst a male from the Aru Islands, Indonesia on 10th April 1872 has a chestnut belly mottled with dark blue and would appear to be a hybrid between the races *pandoo* and *philippensis*. The specimen without data, a male, is of one of the races which exhibit a blue belly.

Zoothera lunulata (Australian Ground Thrush)
Endemic to Australia. Number of races unknown due to recent splitting (formerly considered a race of *Z. dauma* White's Thrush). The two specimens bear no more precise collecting data than "Australia".

Turdus torquatus (Ring Ouzel)
Western Palearctic. Three races. The 11 skins represent a single pullus and both sexes. Six specimens originate from localities in Yorkshire (Teesdale and Spurn) and were collected during the months of April (1), June (2), September (2) and October (1). The other specimens were collected in Sweden (October); Hungary (June); Czechoslovakia (May and undated); the Canary Isles (October) and one from an untraced locality has no date appended. The specimens from Hungary and Czechoslovakia exhibit broad, pale feather-fringes to the body feathers (especially the under tail-coverts) and may belong to the race *alpestris*. The specimen from the Canary Isles, taken at Brolles, Fuerteventura on 13th October 1891, is of interest as the species is but an accidental visitor to these islands.

Turdus merula (Blackbird)
Western and southern Palearctic and, marginally, Oriental (introduced to southeastern Australia and New Zealand). 16 races. 30 skins representing both sexes and all age groups. Of the 25 specimens with data 23 originate from the British Isles; the other two were collected in France on 17th December 1890 and Corsica in January 1891.

Turdus obscurus (Eye-browed Thrush)
Central and eastern Palearctic. Monotypic. The two specimens, both first-year birds, were acquired at an untraced locality; one on 5th October 1883 the other on 12th June (no year appended).

Turdus ruficollis (Black-throated Thrush)
Central Palearctic. Two very distinct races, considered by some authors to be separate species. The single specimen, which belongs to the race *atrogularis,* was collected in northern India on an unrecorded date and had been sexed as female upon dissection.

Turdus naumanni (Dusky Thrush)
Central and eastern Palearctic. Two races. Two specimens, both of which belong to the race *eunomus.* A first-winter bird was taken at Hakodate, Japan on 15th March 1884 and an adult was collected at Foochow, China on 22nd January 1887.

Turdus pilaris (Fieldfare)
Palearctic. Monotypic. The ten specimens represent a single pullus and both sexes (six males and three females). With the exception of the pullus from an untraced locality and a male this is devoid of data; the others were collected in Great Britain (mainly Yorkshire) during the winter months.

Turdus philomelos (Song Thrush)
Western and central Palearctic. Four races. 13 specimens. With the exception of a specimen which lacks locality data the others all originate from Great Britain (mainly Yorkshire).

Turdus iliacus (Redwing)
Palearctic. Two races. Ten skins representing first-year and adult plumages. With the exception of two specimens which lack data and one from the Faeroe Islands on 30th October 1946 the others were collected at British localities during the winter months.

Turdus viscivorus (Mistle Thrush)
Western and central Palearctic. Two or three races. The ten skins represent pullus, juvenile, first-winter and adult plumages. Of the nine specimens with data, seven originate from Yorkshire localities whilst the others are from Essex and Ireland. A bird taken at Cawthorne, Yorkshire on 10th May 1943 is unusual in that its plumage is completely xanthochroic.

SYLVIIDAE

Cettia cetti (Cetti's Warbler)
Southern Palearctic. Five races. With the exception of a specimen devoid of collecting data the other six were acquired in Corsica (3); Turkey (2) and Algeria, Africa (1). The bird from Algeria was figured by C.R.Bree in his book *A History of the Birds of Europe, not observed in the British Isles* (1875).

Locustella fluviatilis (River Warbler)
Central Palearctic. Monotypic. The single specimen was collected in Moscow, Russia on 14th May 1886.

Locustella certhiola (Pallas's Grasshopper Warbler)
Central and eastern Palearctic. At least four races. The single specimen was collected at an untraced locality on 10th June 1873.

Locustella naevia (Grasshopper Warbler)
Western and central Palearctic. Four races. The four specimens originate from Merionethshire, Wales on 27th May 1891 and 3rd July 1892; Lincolnshire on 22nd September 1903 and Yorkshire on an unrecorded date.

Locustella lanceolata (Lanceolated Warbler)
Central and eastern Palearctic. Monotypic. The single specimen was collected in Burma on an unrecorded date.

Acrocephalus paludicola (Aquatic Warbler)
Western Palearctic. Monotypic. The single specimen bears no more precise collecting data than "southern France, 13th October 1897". This record is of interest as the species, although regularly recorded in France as a passage migrant, is not particularly common and adds a hitherto unrecorded occurrence to the French list.

Acrocephalus schoenobaenus (Sedge Warbler)
Western and central Palearctic. Monotypic. With the exception of a specimen from Denmark on 2nd August 1849, the other seven originate from localities in the British Isles and represent both juvenile and adult plumages.

Acrocephalus bistrigiceps (Black-browed Reed Warbler)
Eastern and central China. Monotypic. The two specimens were acquired in Japan in September 1921.

Acrocephalus dumetorum (Blyth's Reed Warbler)
Central Palearctic. Monotypic. The two specimens were collected at Muddapur, India on 21st November 1879 and Moscow, Russia on 29th May (no year appended).

Acrocephalus scirpaceus (Reed Warbler)
Western and central Palearctic. Two races. Of the five specimens, four have decipherable collecting data and originate from; Yorkshire on 9th June 1882; Austria on 12th August 1883; southern France on 19th October 1900 and southern Europe on an unrecorded date.

Acrocephalus palustris (Marsh Warbler)
Western Palearctic. Monotypic. The two specimens were collected in Poland on 18th June 1879 and Gloucestershire on 6th June 1937.

Acrocephalus melanopogon (Moustached Warbler)
Western (patchily) and southern central Palearctic. Three races. Two specimens; one from Turkey on 29th January 1879 the other with indecipherable locality data. This last mentioned bird was figured by C.R.Bree in his book *A History of the Birds of Europe, not observed in the British Isles* (1875).

Acrocephalus arundinaceus (Great Reed Warbler)
Palearctic. Three or four races. Ten specimens. The specimens with data (7) originate from France (1); Malta (1); Italy (2); Hungary (1); Japan (1), whilst the other bears no more precise collecting data than "southern Russia".

Hippolais icterina (Icterine Warbler)
Western Palearctic. Monotypic, although some authors recognize two races. The three specimens originate from Yugoslavia during 1883 and an untraced locality (2) on an unrecorded date.

Hippolais polyglotta (Melodious Warbler)
Southwestern Palearctic. Monotypic. The three specimens were collected in Spain on 2nd May 1874; Italy on 28th May 1895 whilst the other is devoid of data.

Hippolais olivetorum (Olive-tree Warbler)
Confined to the eastern Mediterranean as far east as western Turkey. Monotypic. The single specimen was collected at Smyrna (now Izmir), Turkey on 4th July 1881.

Hippolais pallida (Olivaceous Warbler)
Western and central Palearctic and northern Ethiopian. Five races. The three specimens were collected in Kenya, Africa on 9th November 1910; Egypt, Africa on 8th June 1917 and an untraced locality on 15th June 1877.

Sylvia nisoria (Barred Warbler)
Central Palearctic. Two races. The six specimens originate from Hungary (16th May 1883); Britain (20th October 1916 and 21st October 1927); Spain (6th April 1886) whilst one carries no more precise collecting data than "south Russia" and the other is from an untraced locality on 21st June 1885. The bird from Spain is of interest as the species is but an accidental visitor to that country.

Sylvia hortensis (Orphean Warbler)
Southwestern Palearctic. Three or four races. The five specimens originate from Spain in April 1865; Turkey on 18th May 1883; southern France on 9th September 1895; Greece on 17th April 1899 and Gibraltar on an unrecorded date.

Sylvia borin (Garden Warbler)
Western and central Palearctic. Three races. Of the 19 specimens, eight originate from British localities. The others, apart from an individual from an untraced locality and one lacking data, are from France (2); Germany (1); Italy (5) and the Faeroe Islands (1).

Sylvia atricapilla (Blackcap)
Western Palearctic. Five races. The 13 skins represent all age groups and both sexes. With the exception of four that originate from the British Isles (representing juvenile and males), the others were collected in Morocco (both sexes) and France (juvenile and both sexes).

Sylvia communis (Whitethroat)
Western and central Palearctic. Three or four races. Of the ten specimens only six carry data; a juvenile from Durham on 2nd July 1880, an adult male from Devon on 14th May 1884, an adult male from the Isle of Wight on 29th April 1911 and unageable/sexable birds from Spurn, Yorkshire on 4th September 1882, the Isle of Wight on 10th September 1912 and

Essex on 21st January 1887. This latter bird is a very unusual winter record and a label attached to the skin reads *"Apparently a young bird hatched too late to migrate"*. We now know that this statement is factually incorrect.

Sylvia curruca (Lesser Whitethroat)
Western and central Palearctic. Up to 14 races described. Of the ten specimens, five were taken at British localities during the autumn months; the others originate from southern France on 9th May 1891; Germany on 14th April 1898; Siberia on 25th June 1910 and untraced localities on 26th January 1880 and 6th June 1897.

Sylvia nana (Desert Warbler)
Southwestern Palearctic (patchily in north Africa only) and southern central Palearctic. Three races. The single specimen was collected on the species' wintering grounds at Jodhpur, Rajasthan, India on 29th January 1878.

Sylvia ruppelli (Ruppell's Warbler)
Confined to Greece and western Turkey. Monotypic. The three specimens originate from Turkey on 9th April 1872 (adult male), 16th July 1887 (unageable/sexable) and 8th August 1887 (unageable/sexable).

Sylvia melanocephala (Sardinian Warbler)
Southwestern Palearctic. Up to seven races described. The seven specimens originate from Corsica on 27th December 1890 (first-year female and male) and 9th January 1891 (female); France on 24th January 1886 (male); Italy in May 1906 (male); Berlin, Germany on Ist June (no year is appended) and a male with no more precise collecting data than "Europe". The German bird, a male, is of interest as the species is but of an accidental occurrence in that country. The data label attached to this bird bears the inscription *"Sylvia atricapilla* Black Cap". Although the word *atricapilla* has been crossed out at a later date, a new determination was not added.

Sylvia melanothorax (Cyprus Warbler)
Endemic to Cyprus. Monotypic. The nine specimens originate from two localities on Cyprus; seven of these being males collected in spring 1905.

Sylvia cantillans (Subalpine Warbler)
Southwestern Palearctic. Three races. The two specimens, both males, originate from Tuscany, Italy in 1901 whilst the other has no more precise collecting data than "Europe, 3rd May 1891".

Sylvia undata (Dartford Warbler)
Southwestern Palearctic. Up to five races described. The four specimens, all males, were collected at Churt, Surrey in May 1861 and on 18th November 1861; New Forest, Hampshire on 25th June 1936 and Hyeres, France on 18th December 1890. The Hampshire bird is very dark in appearance and can be attributed to the race *dartfordiensis*.

Phylloscopus trochilus (Willow Warbler)
Palearctic. Three races. 29 specimens. Of the 25 specimens with decipherable data, 23 originate from British localities. The other two were collected in Norway on 2nd July 1888 and Poland in May 1889.

Phylloscopus collybita (Chiffchaff)
Palearctic. Up to eight races described. 15 specimens. With the exception of six specimens from the British Isles, the others originate from Sweden (2), Corsica (2), the Canary Isles (3), Egypt, Africa (1) and India (1).

Phylloscopus sibilatrix (Wood Warbler)
Western Palearctic. Monotypic. 11 specimens. With the exception of three specimens from untraced foreign localities, the others originate from England (3) and Wales (5).

Phylloscopus affinis (Tickell's Warbler)
Extreme southeastern Palearctic and northern Oriental. Two races. The two specimens, both sexed as female upon dissection, were collected in Kashmir, India; one on 15th July 1908, the other some five days later. Both specimens belong to the nominate race. On the label attached to the specimen from 15th July it is noted that the bird had a nest of three eggs and the nest was made of dry grasses lined with crows' feathers (species unknown). The specimen from 20th July bears a similar inscription, but in addition to the nest containing three eggs there was also that of a cuckoo (species unknown) and the nest was lined with Choughs' feathers (either *Pyrrhocorax pyrrhocorax* Chough or *P. graculus* Alpine Chough).

Phylloscopus fuscatus (Dusky Warbler)
Central and eastern Palearctic. Two races. The single specimen was acquired in Upper Assam, India on 22nd October 1903.

69

Phylloscopus inornatus (Yellow-browed Warbler)
Eastern Palearctic. Three races, although some authors now regard these as separate species (*P. inornatus, P. humei* and *P. mandellii*). The two specimens, which originate from Muddapur, India (7th November 1879) and Ubekskaya, Russia (18th August 1930), almost certainly belong to the race (or species) *humei.*

Phylloscopus borealis (Arctic Warbler)
Palearctic and northwestern Nearctic. Up to seven races described. Six of the seven specimens were taken in Japan in November 1921, the other is from an untraced locality on an unrecorded date.

Phylloscopus sp.
The single problematical specimen was taken at an untraced locality on 22nd March 1905. Superficially the skin resembles *P. inornatus* (Yellow-browed Warbler) but the tail feathers (only a few outer feathers remain) show a white patterning suggestive of *P. pulcher* (Orange-barred Warbler). There are two possibilities to the identity of this bird:
1) It may be a hybrid between *P.* (*inornatus/humei*) *mandellii* and *P. pulcher.*
2) It is possible that the skin is that of *P.* (*inornatus/humei*) *mandellii* to which someone has attached the outer tail feathers of *P. pulcher.*
The use of DNA fingerprinting may be the only way to solve this intriguing problem!

Seicercus burkii (Black-browed Flycatcher Warbler)
Oriental (Himalayas and western Burma only). Up to eight races described. The two specimens were taken in India on 8th October 1870 and at an untraced locality in August 1874.

Seicercus xanthoschista (Grey-headed Flycatcher Warbler)
Oriental (Himalayas and western Burma only). Four races. The single specimen was collected at Naini Tal, Uttar Pradesh, India on 15th May 1909.

Regulus satrapa (Golden-crowned Kinglet)
Nearctic. Three races. The single specimen, a female, was collected on Constitution Island, New York, U.S.A. on 17th November 1877.

Regulus calendula (Ruby-crowned Kinglet)
Nearctic. Four races. The two specimens, both males, were taken near Grand Crossing, Illinois, U.S.A. on 19th April 1884.

Regulus regulus (Goldcrest)
Palearctic. Up to 14 races described. 22 skins representing both sexes. The 21 specimens with data originate from British localities with the exception of three from France and singles from Switzerland, Romania and the Canary Isles. The specimen collected on Tenerife, Canary Isles on 15th February 1887 belongs to the race *teneriffae*.

Regulus ignicapillus (Firecrest)
Western Palearctic. Three or four races. Nine skins representing both sexes. The six specimens with decipherable data originate from France (male, female and unsexable), Corsica (female), Spain (female) and the British Isles (male).

Cisticola juncidis (Zitting Cisticola or Fan-tailed Warbler)
Southwestern Palearctic, Ethiopian and Oriental. 18 races. The two specimens originate from as far afield as Muddapur, India (19th June 1880) and southern Spain (10th July 1884).

MUSCICAPIDAE

Ficedula hypoleuca (Pied Flycatcher)
Western Palearctic. Four or five races. The 19 skins represent the plumages of pullus, first-year and adults of both sexes. Of the 18 specimens with locality data, 15 originate from the British Isles whilst others were taken in Norway (two males) and Malta (male).

Ficedula albicollis (Collared Flycatcher)
Central western Palearctic. Monotypic. Four specimens; two adult males and an adult female from Germany and an adult male from Gotland, Sweden. With the exception of the Swedish bird that was collected in May, the others were taken in April.

Ficedula zanthopygia (Yellow-rumped Flycatcher)
Extreme southeastern Palearctic and northeastern Oriental. Monotypic. The single specimen, a female, was taken in Ussuri Febiet, China on 21st May 1887.

Ficedula parva (Red-breasted Flycatcher)
Palearctic. Two or three races. Five specimens (four adults and a first-year). The adults were acquired in Moscow, Russia on 30th April 1891; southern Russia on an unrecorded date; Yugoslavia on an unrecorded

71

date and Germany on the 14th May of an unrecorded year. The first-year was taken in Romania on 5th September 1900.

Muscicapa striata (Spotted Flycatcher)
Western and central Palearctic. Up to nine races described. The eight specimens represent pullus, first-year and adult plumages. With the exception of an adult from Norway in July (no year appended) the others were acquired at British localities.

AEGITHALIDAE

Aegithalos caudatus (Long-tailed Tit)
Palearctic. 19 races. 22 skins representing the plumages of juvenile and full-grown birds. With the exception of a specimen from an untraced locality the others originate from the British Isles (3); the Pyrenees, France (6); Gibraltar (1); Corsica (3); Sweden (2); Norway (1) and Hakodati, Japan (5). The specimens from Japan belong to one of the white-headed races.

REMIZIDAE

Remiz pendulinus (Penduline Tit)
Western and central Palearctic. Up to 16 races described. Three skins representing the plumages of juvenile (1) and full-grown birds (2). The juvenile was taken in Turkey on the 19th July of an unrecorded year whilst the others, both sexed as male upon dissection, were acquired in the Lower Volga, Kalmytskaya, Russia on 30th April of an unrecorded year and Sarepta, Kalmytskaya, Russia on 8th May 1888.

PARIDAE

Parus palustris (Marsh Tit)
Western and eastern Palearctic and, marginally, Oriental. Up to 16 races described. Six specimens. With the exception of a specimen from Gothenburg, Sweden (23rd March 1909), the others originate from Yorkshire (December 1883 (2) and December 1927); Devon (January 1886); Surrey (December 1947).

Parus lugubris (Sombre Tit)
Western Palearctic (southeastern only). Six or seven races. The two specimens, a juvenile and a full-grown individual, were taken in Turkey (juvenile on 4th July 1881 and full-grown on 12th February 1879).

Parus montanus (Willow Tit)
Palearctic. Up to 14 races described. Ten specimens. With the exception of a specimen from an untraced locality the others originate from Yorkshire (3); Sweden (5) and Norway (1).

Parus atricapillus (Black-capped Chickadee)
Nearctic. Seven races. The three specimens were taken in Ontario, Canada (18th January 1888 and 25th January 1889) and an untraced locality (10th November 1893).

Parus ater (Coal Tit)
Palearctic and, marginally, Oriental. Up to 20 races described. 12 specimens. The specimens with locality data (11) originate from the British Isles (5); France (3); Germany (1); Lake Baikal, Buryatskaya, Russia (1) and an untraced locality (1).

Parus cristatus (Crested Tit)
Western Palearctic. Up to nine races described. The seven specimens were collected in Sweden (4); Norway (1); Germany (1) and the Pyrenees, France (1).

Parus major (Great Tit)
Palearctic and Oriental. At least 33 races. Of the 17 specimens, which represent both sexes and juvenile plumages, 12 were acquired at British localities whilst the others originate from the Pyrenees, France (male and female); Corsica (female); Norway (juvenile) and Sweden (female).

Parus caeruleus (Blue Tit)
Western Palearctic. 15 or 16 races. The 15 skins represent the plumages of first-years and adults. Of the 14 specimens with data, ten originate from the British Isles whilst the others were collected on Tenerife, Canary Isles (3) and the Pyrenees, France (1). The three specimens from Tenerife can presumably be ascribed to the race *teneriffae* and the specimen without locality data is exhibiting characteristics indicative of the race *ultramarinus*.

SITTIDAE

Sitta europaea (Nuthatch)
Palearctic and Oriental. Up to 28 races described. The nine skins represent the plumages of juvenile and both sexes. With the exception of two

specimens that are devoid of locality data the others were acquired in England (4); Sweden (1) and the Pyrenees, France (2).

Sitta kruperi (Kruper's Nuthatch)
Restricted to Turkey and the Caucasus. Monotypic. The single specimen, said to be male upon dissection, was collected at Smyrna (now Izmir), Turkey on 8th November 1882.

Sitta neumayer (Rock Nuthatch)
Western Palearctic (southeastern only). Five or six races. Five specimens. With the exception of two specimens that are devoid of locality data (collected on 22nd May 1883 and 8th June 1884), the others were acquired in Turkey on 14th June 1881, 22nd May 1884 and 4th May of an unrecorded year.

TICHODROMADIDAE

Tichadroma muraria (Wallcreeper)
Southern Palearctic and northern Oriental. Two races. The two specimens were acquired in the Alpes de Savoie, France on 20th October 1870 and the Hautes Pyrenees, France in May 1883. This last mentioned specimen is showing signs of summer plumage in the throatal region.

CERTHIIDAE

Certhia familiaris (Treecreeper)
Holarctic and, marginally, Oriental. Approximately 25 races described but variation is slight and clinal (apart from isolated races). This species is only represented by two specimens (1 Swedish, 1 Corsican). This general lack of specimens may be due to the fact that the species is reluctant to place itself in a position to be shot at!

EMBERIZIDAE

Miliaria calandra (Corn Bunting)
Western Palearctic. Two races. Nine specimens. With the exception of specimens from Turkey in April 1866 and Ubekskaya, Russia on 27th May 1930, the others originate from English localities.

Emberiza citrinella (Yellowhammer)
Western Palearctic. Three races. The 17 skins represent both sexes and juvenile plumages. With the exception of specimens from the Pyrenees,

France (adult female on 20th June 1883); Norway (juvenile in July 1888) and three which are devoid of collecting data, the other 12 were acquired in the British Isles.

Emberiza leucocephala (Pine Bunting)
Central and eastern Palearctic. Two races. The single specimen, a female, was collected in Ubekskaya, Russia on 27th June 1884.

Emberiza citrinella (Yellowhammer) X *Emberiza leucocephala* (Pine Bunting)
The single specimen, an adult male, was collected at Arkhangel'sk, Russia on 1st July 1872. Although appearing as *E. citrinella*, the specimen has reddish malar region markings and is, therefore, showing evidence of hybridisation with *E. leucocephala*. The races which regularly hybridise in an overlap zone are *E. c. erythrogenys* and *E. l. leucocephala*.

Emberiza cia (Rock Bunting)
Southern Palearctic. 11 races. The eight specimens originate from the Pyrenees, France (four males and a female) and Kashmir, India (two juveniles and a female-type). The wide geographical provenance of the specimens ensures that at least two races are available.

Emberiza cioides (Siberian Meadow Bunting)
Eastern Palearctic and northeastern Oriental. Five races. The five specimens, three males and two females, were acquired at Foochow, China on 2nd January 1885 (male) and 1st February 1885 (male); Japan in 1893 (female) and Tokyo, Japan on unrecorded dates (male and female). The two females had previously been examined by D.T. Lees-Smith and ascribed to the races *cioides* and *ciopsis*.

Emberiza hortulana (Ortolan Bunting)
Western Palearctic. Monotypic. The six skins represent both sexes and juvenile plumages. With the exception of a male that is devoid of locality data (collected on 16th May 1878) the others originate from Switzerland in 1876 (male); Sarepta, Kalmytskaya, Russia on 1st July 1882 (female) and Turkey on 29th April 1881 (female), 12th August 1887 (juvenile) and 17th April 1894 (male).

Emberiza caesia (Cretzschmar's Bunting)
Southwestern Palearctic. Monotypic. The two specimens, both juveniles, were collected at Smyrna (now Izmir), Turkey in July 1875 and in Greece on 12th June 1883.

Emberiza cirlus (Cirl Bunting)
Western Palearctic. Two races. The ten specimens, five male and five females, originate from southern England (female); France (male); Corsica (male and two females); Spain (female); Italy (male); Switzerland (male) and Greece (male and female).

Emberiza fucata (Chestnut-eared Bunting)
Oriental. Three races. The single specimen, a male, was collected in Japan on an unrecorded date. The skin had previously been examined by D.T. Lees-Smith and ascribed to the nominate race.

Emberiza pusilla (Little Bunting)
Northern Palearctic. Monotypic. Five specimens. With the exception of a specimen that lacks collecting data, the others were acquired in West Bengal, India on 2nd March 1879 and 16th December 1879; Arkhangel'sk, Russia on 18th June 1889 and an untraced locality on 22nd July 1872.

Emberiza rustica (Rustic Bunting)
Northern Palearctic. Two races. The nine specimens originate from Arkhangel'sk, Russia (4th July 1872 and undated); China (5th October 1883) and Japan (22nd October 1884, 2nd November 1884 and four undated).

Emberiza aureola (Yellow-breasted Bunting)
Northern Palearctic. Monotypic. The four specimens, which represent juvenile, adult male, second-year male and adult female, were collected at Arkhangel'sk, Russia on 29th August 1883 (juvenile) and China on 9th May 1884 (adult male), 14th May 1884 (adult female) and 26th May 1885 (second-year male).

Emberiza melanocephala (Black-headed Bunting)
Southwestern Palearctic. Monotypic. Seven skins (five adult males and two adult females). The six specimens with data originate from Turkey (male); Italy (female); Greece (male and female) and Russia (two males).

Emberiza bruniceps (Red-headed Bunting)
Southern central Palearctic. Monotypic. The two specimens, adults of both sexes, were collected in Kazakhstan, Russia; the female on 2nd May of an unrecorded year and the male on 3rd May of an unrecorded year.

Emberiza sulphurata (Japanese Yellow Bunting)
Restricted to central Japan. Monotypic. The two specimens, both females, were collected at Minamiajumi Shinano, Japan in October 1920.

Emberiza spodocephala (Black-faced Bunting)
Southeastern Palearctic and, marginally, northern Oriental. Three races. The two specimens, both males, were collected in Japan in 1893. Both skins had previously been examined by D.T. Lees-Smith and ascribed to the race *personata*.

Emberiza schoeniclus (Reed Bunting)
Palearctic (save for the northeast). Up to 15 races described. The 27 skins represent both sexes and all age groups (with the exception of pullus). The specimens with data (23) originate from the British Isles (20); Ireland (1); Pyrenees, France (1) and Ubekskaya, Russia (1).

Calcarius lapponicus (Lapland Bunting)
Holarctic. Three races. Seven specimens. The specimens with locality data (5) originate from Ust Tsilma, Russia (summer plumaged male on 25th May 1875); Norway (summer plumaged male on 20th June 1883 and first-year on 26th June 1883); Belgium (summer plumaged male on an unrecorded date) and Greenland (first-year on an unrecorded date). The two skins lacking locality data are an adult male in summer plumage and an adult in winter plumage.

Plectrophenax nivalis (Snow Bunting)
Holarctic. Four races. The 27 skins represent the plumages of juvenile and both sexes (summer and winter). The specimens with data (21) originate from the British Isles (13); the Faeroe Islands (1); Norway (2); Heligoland, Germany (1); Kurile Islands, Sakhalinskaya Oblast, Russia (1); western Russia (2) and Siberia (1).

FRINGILLIDAE

Fringilla coelebs (Chaffinch)
Western and central Palearctic (introduced elsewhere). 14 races. The 19 specimens, representing both sexes, were acquired in the British Isles (six males and six females); Ireland (three males); France (female); Morocco, Africa (male) and the Canary Isles (male and female).

Fringilla teydea (Blue Chaffinch)
Endemic to the western Canary Isles. Two races. The two specimens, a male and female, were taken on Tenerife, Canary Isles; the female on 10th February 1889 and the male some eight days later.

Fringilla montifringilla (Brambling)
Northern Palearctic. Monotypic. 25 skins representing both sexes. The specimens with data (20) originate from the British Isles with the exception of a single from Montpellier, France in January 1886 and six from the species' breeding grounds in Norway during May 1924 and May 1925.

Serinus pusillus (Red-fronted Serin)
Patchily distributed between Turkey in the west and northwest China in the east. Monotypic. The single specimen, sexed as male upon dissection, was taken at Taschkend, Uzbekskaya, Russia on 16th May 1876.

Serinus serinus (Serin)
Southwestern Palearctic. Monotypic. The seven specimens (three males and four females) originate from France (two males and a female); Hungary (male); Corsica (female); Malta (female) and Morocco, Africa (female).

Serinus canaria (Canary)
Restricted to the Azores, Madeira and the western Canary Isles. Monotypic. Six specimens. Four, of which three had been sexed as male upon dissection, originate from Tenerife, Canary Isles whilst the other two were acquired on the Azores.

Serinus citrinella (Citril Finch)
Central western Palearctic. Two races. The seven specimens originate from the Pyrenees, France on 7th June 1883 (3) and 3rd June 1896; Andorra, Spain on 21st May 1889 and Corsica in January 1906 and 29th January 1911.

Carduelis chloris (Greenfinch)
Western Palearctic (introduced elsewhere). Four races. 12 skins representing both sexes and all age groups. The specimens originate from British localities with the exception of a male from Corsica (29th December 1890) and two that are devoid of collecting data.

Carduelis sinica (Oriental Greenfinch)
Southeastern Palearctic and, marginally, northeastern Oriental. Six races. The two specimens, both males, were taken in Tokyo, Japan on an unrecorded date.

Carduelis spinus (Siskin)
Discontinuous, with populations in western and eastern Palearctic. Monotypic. 14 specimens representing both sexes. The 13 specimens with data were acquired at localities in the British Isles.

Carduelis carduelis (Goldfinch)
Western and central Palearctic and, marginally, Oriental (introduced elsewhere). Up to 19 races described. The 15 skins represent juvenile and adult plumages. The 14 specimens with locality data originate from the British Isles (8); Ireland (1); France (1); Italy (1); Austria (1); Sarepta, Russia (1) and an untraced locality (1). This last mentioned bird belongs to one of the grey-headed races.

Carduelis carduelis (Goldfinch) X *Serinus canaria* (Canary)
The single hybrid between these two species had been bred in captivity at Church Fenton, Yorkshire on an unrecorded date. The *S. canaria* was of a form kept by aviculturists and not of pure stock.

Carduelis flammea (Redpoll)
Holarctic. Four races. 44 specimens. The 41 skins with locality data originate from the British Isles (22); Norway (8); Denmark (1); Sweden (1); an unrecorded country in Lapland (1); Arkhangel'sk, Russia (3); Greenland (4) and Ontario, Canada (1). An adult female from an unrecorded locality on 13th June (no year appended) had been shot on a nest containing four eggs (the whereabouts of the nest and eggs, if collected, is unknown).

Carduelis hornemanni (Arctic Redpoll)
Holarctic. Two races. The five skins can be clearly divided into the two races. Examples of the nominate race were taken at Fort Chimo, Quebec, Canada (5th December 1882 and 28th April 1883) and the Faeroe Islands (30th April 1945). The others, two males belonging to the race *exilipes*, were acquired at Hammerfest, Norway (2nd July 1888) and Scarborough, Yorkshire (18th December 1925). The label attached to the Faeroe Island bird, which was collected at Miklidalur, Kallsoy, indicates that this was the first record of this species for the islands. The Yorkshire bird was picked up in an injured condition and, although there was confusion over taxonomy (and still is), provided the county with its seventh authentic record.

Carduelis flavirostris (Twite)
Discontinuous, with populations in northwestern Europe and central and eastern Asia. Eight races. The 18 specimens represent the plumages of both sexes. The specimens with data (17) originate from the British Isles (mainly the Yorkshire coast but also northern Scotland) and Ubekskaya, Russia. The Russian birds (3), collected in August 1929, are a pale buffish colour and may belong to the race *montanella.*

Carduelis cannabina (Linnet)
Palearctic and, marginally, Oriental. Five or six races. Six specimens (four males and two females). With the exception of a female from the Pyrenees, France (15th January 1886) the others originate from the British Isles.

Leucosticte nemoricola (Hodgson's Mountain Finch)
Northern Pakistan, Himalayas to western China. Two races. Nine specimens. With the exception of a bird from an unrecorded locality (collected on 5th June 1906), the others originate from Kashmir, India (25th March 1904, 20th October 1904 and 2nd August 1908); Mimachal Mandi Pradesh, India (three on 20th March 1911) and two untraced localities (one on 21st April 1930 the other on an unrecorded date).

Leucosticte brandti (Brandt's Mountain Finch)
Central Asia and northern Pakistan to northwestern China and Mongolia. At least five races. The two specimens, both adults, were taken at an untraced locality on 25th July 1929.

Leucosticte arctoa (Rosy Finch)
Eastern Palearctic and western Nearctic. 14 races. The two specimens, a male and a female-type, originate from Japan on an unrecorded date.

Bucanetes githaginea (Trumpeter Finch)
Southern Palearctic, northern Ethiopian and, marginally, Oriental. Four races. Three specimens; a male and female-type were acquired on Fuerteventura, Canary Isles in March 1889 whilst the other, a male, is devoid of data.

Rhodopechys obsoleta (Desert Finch)
Southern central Palearctic. Monotypic. The single specimen was acquired in Jordan on the 20th January of an unrecorded year.

Uragus sibiricus (Long-tailed Rosefinch)
Eastern Palearctic and northern Oriental. Five races. The two specimens, both males of the race *sanquinolentus,* were taken in Tokyo, Japan on an unrecorded date.

Carpodacus erythrinus (Scarlet Rosefinch)
Palearctic and, marginally, northern Oriental. Five races. The ten specimens,of which three are in adult male plumage, were acquired in Azerbaydzhan, Russia (female-type on 30th March of an unrecorded year and a male on an unrecorded date); Kalmytskaya, Russia (two males on unrecorded dates); Arkhangel'sk, Russia (female-type on 16th May 1879); Kazakhskaya, Russia (two female-types on 13th July 1929) and two untraced localities (female-types on 9th May 1907, 13th July 1908 and 10th May 1909).

Carpodacus vinaceus (Vinaceous Rosefinch)
Southeastern Palearctic and northern Oriental. Two races. The single specimen, a female, was taken at Taipaishang, China on 19th September 1905.

Carpodacus roseus (Pallas's Rosefinch)
Eastern Palearctic. Monotypic. The three specimens, all males, were acquired in Japan; two carry no collecting date and the other simply indicates the year "1893".

Pinicola enucleator (Pine Grosbeak)
Holarctic. Ten or 11 races. The 15 specimens represent all plumage types with the exception of pullus. The specimens with data (10) all originate from Sweden (two juveniles, a first-year male, five adult males, an adult female and a female-type). One of the adult males had been shot on a nest containing four eggs on 6th June 1862 (the whereabouts of the nest and eggs, if collected, is unknown). This record is of interest as Harrison's book *A Field Guide to the Nests, Eggs and Nestlings of British and European Birds* (1987) states that only the female of this species incubates.

Loxia pytyopsittacus (Parrot Crossbill)
Western Palearctic (northern only). Monotypic. The 17 skins represent the plumages of both sexes and juvenile. The 16 specimens with data originate from Sweden (two juveniles, four males and three females); Norway (two males and a female); Latvia, (male) and Britain (male and two females). The last mentioned birds, taken near Colchester, Essex on 21st February 1862, are of interest as the species is but an accidental visitor to these shores (although they are probably overlooked). In

81

addition, five recently fledged juveniles labelled as *L. pytyopsittacus* (two from Sweden; the others devoid of locality data), may well belong to this species. The undeveloped mandibles, however, preclude positive identification.

Loxia scotica (Scottish Crossbill)
Endemic to northern Scotland. Monotypic. There is a single male specimen of what is now Britain's only indigenous bird species. The specimen concerned was taken at Dores, near Inverness, Highland, Scotland on 19th April 1925. Additionally, a male from Dalwhinnie, Highland, Scotland may well belong to this species, but the biometrics of the specimen's bill (the only way positively to separate this species from *L. curvirostra*) give no indication as to its true identity.

Loxia curvirostra (Crossbill)
Holarctic and, marginally, Oriental. 20 races. The 18 specimens, representing the plumages of juvenile and both sexes, originate from the British Isles (juvenile, four males and two females); Ireland (male); Sweden (three males and a female); Andorra, Spain (two females); Arkhangel'sk, Russia (female); Tokyo, Japan (two males) and a male from the Carpathians (country not indicated). The wide geographical provenance of the specimens ensures that a good number of races are available.

Loxia leucoptera (Two-barred Crossbill)
Holarctic. Three races. The five skins represent both sexes and juvenile plumage. The only specimens with collecting data were acquired at Arkhangel'sk, Russia (a juvenile on 2nd March 1884 and a male in the September of an unrecorded year) and the other, a juvenile, bears no more precise locality data than "North America". The geographical provenance of the small sample does, however, ensure that two races are present.

Pyrrhula erythaca (Beavan's Bullfinch)
Extreme southeastern Palearctic and northeastern Oriental. Three races. The five specimens, three males and two females, were taken at Taipaishang, China in July 1905.

Pyrrhula aurantiaca (Orange Bullfinch)
Restricted to northern Pakistan, Kashmir and northwestern India. Monotypic. The single specimen, a female, was acquired in Kashmir, India on 28th December 1907.

Pyrrhula pyrhhula (Bullfinch)
Palearctic. Ten races. 22 skins representing both sexes and juvenile plumages. The 20 specimens with locality data originate from the British Isles (one juvenile, four males and three females); France (female); Norway (two males and a female); Sweden (two males); western Russia (three males) and Japan (three males). The birds from Japan (on unrecorded dates) belong to the very distinctive race *griseoventris*. The distinctiveness of this race has led some authors into regarding it as a full species. A specimen, sexed as male upon dissection, that had been kept in confinement at an unrecorded locality is melanic, being predominantly black on the upperparts (glossy blue-black on the crown, tertials and tail) and peppered with black on the underparts. The bases of primaries five and six (numbered ascendently) are smokey-white.

Coccothraustes coccothraustes (Hawfinch)
Palearctic. Five races. The 11 specimens, representing the plumages of juvenile (both sexes) and adults of both sexes, originate from English localities with the exception of a juvenile male from Sweden on 16th July 1874; a female from France on 6th January 1891; a female from the Kurile Islands, Sakhalinskaya, Russia on an unrecorded date; a male from an untraced locality on 22nd May 1874 and a male from an unrecorded locality in 1879.

ESTRILDIDAE

Amandava subflava (Zebra Waxbill)
Sub-Saharan Africa. Two races. The single male specimen is without data and may well have originated from captive stock.

PASSERIDAE

Passer domesticus (House Sparrow)
Cosmopolitan. 11 races. 25 skins representing both sexes. With the exception of three specimens that are devoid of locality data, the others originate from England (15); France (3); Italy (2); Hungary (1) and Turkey (1).

Passer hispaniolensis (Spanish Sparrow)
Southern Palearctic. Two or three races. Five specimens (all males). With the exception of a specimen that is devoid of locality data the others originate from Greece (1); Turkey (1) and Egypt, Africa (2).

Passer domesticus / Passer hispaniolensis (House / Spanish Sparrow)
The skin of a female from Egypt, Africa on 24th October of an unrecorded year is labelled *Passer hispaniolensis hispaniolensis*. Unfortunately, however, there is nothing to indicate that it belongs to this species and the bird may be *P. domesticus*.

Passer montanus (Tree Sparrow)
Palearctic and Oriental. Seven races. 34 specimens. With the exception of a skin that is devoid of locality data, the others originate from England (16); Italy (2); Assam, India (1); Taiwan, China (4); China (2); Japan (4); the Kurile Islands, Sakhalinskaya Oblast, Russia (1) whilst three others were acquired at two untraced localities.

Passer ammodendri (Saxaul Sparrow)
Discontinuously between central Asia in the west and Mongolia in the east. Three or four races. The six skins, four male and two females, were acquired in Ubekskaya, Russia.

Passer moabiticus (Dead Sea Sparrow)
Patchily distributed between Cyprus in the west and western Afghanistan in the east. Two races. The single specimen, a female, bears no more precise collecting data than "Central Asia, 2nd June 1901".

Passer rutilans (Cinnamon Sparrow)
Extreme southeastern Palearctic and eastern Oriental. Three races. The two specimens, both males, were collected in Japan on unrecorded dates.

Petronia petronia (Rock Sparrow)
Southern Palearctic. Seven races. The six specimens, all adults, were acquired in France (3); Turkey (2) and England (1). The English bird bears a label which simply states that it "died at Lilford". This species was not added to the British List until 1981 (Norfolk) and, on the surface, this would appear to be an earlier record (although no date is appended), there being a Lilford in both Greater Manchester and Northampton. This is not considered to be the case, however, and the skin is probably not referable to that of a genuine vagrant. A far more feasible explanation would appear to be that the bird had died in Lord Lilford's menagerie at Lilford Hall, near Oundle, Northampton and a skin was prepared from the corpse!

Montifringilla nivalis (Snow Finch)
Patchily distributed across the southern Palearctic and northern Oriental. Four races. The single specimen was collected in Switzerland on an unrecorded date.

STURNIDAE

Sturnus vulgaris (Starling)
Western and central Palearctic. 11 races. The 54 skins represent the plumages of all age groups (excluding pullus) and both sexes. The specimens were acquired at English localities with the exception of specimens from Ireland (1); Shetland, Scotland (1); the Faeroe Islands (10); the Altai region of Russia (1) and a single that is devoid of collecting data.

Sturnus unicolor (Spotless Starling)
Western Palearctic (countries bordering the western Mediterranean and a few Mediterranean islands). Monotypic. The three specimens originate from Gibraltar (adult on 9th May 1872); southern Spain (first-year on 6th December 1884) and southern France (adult on 5th June 1886).

Sturnus roseus (Rose-coloured Starling)
Southern central Palearctic. Monotypic. The five specimens were collected in Turkey (juvenile on 20th July 1882); Syria (second-year on 20th May 1883); Bulgaria (two adults on 2nd June 1866) and southern France (adult in the June of an unrecorded year). The French bird is of interest as this species is but an accidental visitor to that country.

Acridotheres tristis (Common Mynah)
Oriental. Three races. The single specimen, which had been sexed as male upon dissection, was taken at Banks, Southport, Lancashire in January 1938. This resident species occurs no further west than southeastern Iran and it is obvious that this is nothing more than an escape from captivity.

ORIOLIDAE

Oriolus oriolus (Golden Oriole)
Western and central Palearctic and northern Oriental. Two races. Six specimens. With the exception of an adult male that is devoid of locality data, the others were acquired in the Pyrenees, France (adult male); southern Spain (first-year female); Hungary (first-year male) and Bulgaria (first-year male and adult male).

PARADISAEIDAE

Ptiloris victoriae (Queen Victoria Riflebird)
Endemic to northeastern Queensland, Australia. Monotypic. The single female was collected in Queensland, Australia on the 29th September of an unrecorded year.

Paradisaea raggiana (Raggiana Bird of Paradise)
Endemic to Papua New Guinea. Five races. The single adult male bears no collecting data.

CORVIDAE

Garrulus glandarius (Jay)
Palearctic and northern Oriental. Up to 36 races described. 16 specimens. Of the 13 specimens with locality data, seven originate from the British Isles whilst the others were acquired in Belgium (1); Corsica (1); Germany (1); Syria (1) and China (2). A specimen from Buttercrambe Wood, near York, Yorkshire on 3rd July 1945 is showing albinistic tendencies, being white and grey throughout (the bird was sexed as female upon dissection). A Yorkshire specimen has, on one side of the label, the inscription "*Garrulus glandarius glandarius*. Bolton Percy, Yorks. 17.3.1947". The other side of the label indicates the date as 17.3.1948 and notes that this is the "First record for Yorkshire. see Brit. Birds 40: (1947)." The data has been wrongly transposed and should read 17.3.1947 and, although the British Birds page number is not indicated, it is now known to be page 211. This was the first recorded instance of the Continental race in Yorkshire, the bird being very grey on the mantle and lacking any deep vinous tinge. The bird was sexed as female upon dissection, having ovaries the size of pin-heads. The wide geographical provenance of the specimens ensures that a good number of races are available.

Cyanopica cyana (Azure-winged Magpie)
Disjunct, with populations in the southwestern Palearctic (Iberian Peninsula only) and southeastern Palearctic and northern Oriental. Nine races. The single specimen is devoid of locality data but was acquired on 3rd April 1880.

Pica pica (Magpie)
Holarctic, Oriental and Ethiopian (Saudi Arabia only). 13 races. The 12 specimens, which represent juvenile, first-year and adult plumages, were acquired at British localities with the exception of a first-winter from Guernsey, Channel Islands and three which are devoid of collecting data.

Nucrifraga caryocatactes (Nutcracker)
Palearctic. Ten races. The two specimens, both of which are first-year examples of the nominate race, were taken in Switzerland in September 1896. Upon dissection the specimens had been sexed as a male and a female.

Pyrrhocorax pyrrhocorax (Chough)
Southern Palearctic and northeastern Ethiopian. Seven or eight races. The five specimens, of which four carry collecting data, originate from Scotland (3) and County Donegal, Ireland (1).

Pyrrhocorax graculus (Alpine Chough)
Southern Palearctic and northern Oriental. Two races. Two specimens: an adult from southern France on an unrecorded date and a juvenile from an untraced locality on an unrecorded date.

Corvus monedula (Jackdaw)
Western and central Palearctic. Four races. The 25 specimens were collected at British localities with the exception of specimens from Sweden (4) and Norway (1). A specimen taken near Hull, Yorkshire on 10th October 1941 is a leucistic individual, being predominantly grey in plumage.

Corvus frugilegus (Rook)
Palearctic. Two races. The six specimens, five of which carry data, were taken at British localities and represent juvenile, first-year and adult plumages.

Corvus corone (Carrion Crow)
Palearctic and, marginally, Oriental. Six races. The 37 skins represent the races *corone* and *cornix* along with a hybrid between these two races. The 13 specimens belonging to the nominate race were acquired in Yorkshire (7); Lancashire (2); Northamptonshire (1) and Glasgow, Scotland (1) whilst the other two are devoid of locality data. The origins of the 23 specimens belonging to the race *cornix* are Yorkshire (9); Nolsoy hin, Faeroe Islands (7); Ireland (2); Scotland (2) whilst the other three are lacking locality data. The hybrid was acquired in Scotland.

Corvus corone (Carrion Crow) X *Corvus corax* (Raven)
The single specimen, which exhibits characters indicative of both *C. c. corone* and *C. corax*, was collected on the Isle of Mull, Scotland on 6th January 1947.

Corvus torquatus (Collared Crow)
Eastern Oriental. Monotypic. The single specimen was acquired at an untraced locality on 28th December 1896. The bird had been sexed as female upon dissection.

Corvus corax (Raven)
Holarctic and northwest Neotropical. Seven or eight races. The five specimens, four of which bear data, were collected in Tangier, Morocco, Africa on 7th June 1893; Nolsoy hin, Faeroe Islands on 21st March 1946 and 27th March 1946 and the Isle of Islay, Scotland on 5th March 1948.

BIRD MOUNTS

With few exceptions the mounts, both cased and uncased, are devoid of collecting data. This was a common practice in the early years, as most collectors simply required to possess a specimen, and were not interested in the scientific aspect of possession. Although this is unfortunate, the mounted specimen does give the scientist an opportunity to study a representative of that particular species, albeit without knowing its geographical provenance.

Bird mounts, unlike skins, are positioned so as to look more life-like and, therefore, appear much as they would under natural conditions. Unfortunately, however, some of the divers and grebes are displayed in an upright position that is physically impossible for such a species.

It can be taken that, unless otherwise indicated, no collecting data is appended to the mount.

CASED MOUNTS

The Museum possesses a total of 149 cases which contain 148 species of bird. For the research student and scientist one of the least beneficial properties of cased mounts concerns their availability, as, unlike skins, they cannot be handled and measured.

As mounts were originally intended to be placed on display, the many years of exposure to daylight eventually fades the plumage of the specimens. As a consequence most of the old cased mounts housed at the Yorkshire Museum suffer from this problem.

PODICIPEDIFORMES

PODICIPEDIDAE

Tachybaptus ruficollis (Little Grebe)
Palearctic, Ethiopian and Oriental. At least nine races. One case has a bird in summer plumage. Another case holds four birds; the only bird with data is in winter plumage and was caught alive at Castle Mills Bridge, York, Yorkshire on 12th September 1911.

PROCELLARIIFORMES

PROCELLARIIDAE

Fulmarus glacialis (Fulmar)
Holarctic. At least two races. A single case holds three light phase birds.

Bulweria bulwerii (Bulwer's Petrel)
Islands in the Atlantic and Pacific Oceans. Monotypic. The single case contains two birds. One of the birds is presumed to be the bird found dead by the River Ure at Tanfield, near Ripon, Yorkshire on 8th May 1837. This was the first European example of this oceanic species and Nelson goes into great detail about this bird in his book on *The Birds of Yorkshire* (1907). The specimen was also used by Lord Lilford for the illustration of this species in his book *British Birds*. The other bird was washed up on the beach at Scalby Mills, Scarborough, Yorkshire on 28th February 1908.

Puffinus puffinus (Manx Shearwater)
Western Palearctic (mainly Great Britain), Nearctic (patchily) and Australasian (New Zealand only). Six races, although some authors now regard the five in the Pacific as separate species. Two birds are housed together.

Puffinus mauretanicus (Balearic Shearwater)
Western Palearctic (Balearic Islands only). Monotypic. The single case holds one bird.

HYDROBATIDAE

Hydrobates pelagicus (Storm Petrel)
Western Palearctic (mainly Great Britain and Mediterranean islands). Monotypic. The single case contains two birds.

Oceanodroma leucorhoa (Leach's Petrel)
Northern Atlantic and northern Pacific Oceans. At least four races. Three birds are cased together. Another case holds a single, caught at Harrogate, Yorkshire on 29th October 1952. This last mentioned bird, said to be a female upon dissection, had been found after a series of severe gales that had been the cause of a 'wreck' of this oceanic species.

PELECANIFORMES

PHALACROCORACIDAE

Phalacrocorax aristotelis (Shag)
Western Palearctic. Three races. An adult and an immature are housed together, as are two breeding plumaged adults.

CICONIIFORMES

ARDEIDAE

Botaurus stellaris (Bittern)
Palearctic and Ethiopian (southern Africa only). Two races. The single case holds a bird belonging to the nominate race that was shot at Figham Pastures, Beverley, Yorkshire during the spring of 1906.

Nycticorax nycticorax (Night Heron)
Almost cosmopolitan (save for Australasian). Four races. The single case contains an adult that was shot at Le Pila (a suburb of Sestri Levante, Liguria, Italy) in June or July 1916.

Ardeola ralloides (Squacco Heron).
Western Palearctic and Ethiopian. Monotypic. The single case holds two birds that are in summer plumage. One of the birds is labelled as being taken at Pontefract, Yorkshire and is no doubt the bird acquired by Arthur Strickland some years prior to 1844 at Askern (near Pontefract). The other bird is from Rudston, Yorkshire.

Egretta alba (Great White Egret)
Almost cosmopolitan. At least four races. Two cases house single adults; one in breeding plumage the other in winter plumage.

CICONIIDAE

Ciconia nigra (Black Stork)
Palearctic and Ethiopian (southern Africa only). Monotypic. Two cases house single birds; an adult and an immature. The immature, despite only being labelled " Market Weighton, 1852" is known to have been captured on Market Weighton Common, Yorkshire about the 29th October of that year.

THRESKIORNITHIDAE

Platalea leucorodia (Spoonbill)
Southern Palearctic, Oriental and Ethiopian (northeastern only). An adult in summer plumage and an immature are housed together.

ANSERIFORMES

ANATIDAE

Branta canadensis (Canada Goose)
Nearctic (introduced in western Palearctic). At least 12 races. The single case holds an adult.

Anas penelope (Wigeon)
Northern Palearctic. Monotypic. The single case contains a male.

Aythya marila (Scaup)
Northern Palearctic. At least two races. A male and female are housed together.

Clangula hyemalis (Long-tailed Duck)
Holarctic. Monotypic. The single case houses five birds, two of which are males in winter plumage.

Bucephala clangula (Goldeneye)
Holarctic. Two races. Three birds, an adult male, an immature male and an adult female, are housed together. Another case holds an adult male and a female-type that were shot at Newton on Derwent, Yorkshire in February 1900.

Mergus merganser (Goosander)
Holarctic. Three races. The single case holds a male and female that were shot by Snowden Sleights at East Cottingwith, Yorkshire on 4th February 1903.

FALCONIFORMES

PANDIONIDAE

Pandion haliaetus (Osprey)
Cosmopolitan, though patchy in Ethiopian and Neotropics. Five or six described races. The single case holds one bird.

ACCIPITRIDAE

Milvus milvus (Red Kite)
Western Palearctic. Two races. The single case contains an adult bird that was shot near Guisborough, Yorkshire on an unrecorded date. The bird is displayed standing over its next meal, a Grey Partridge *Perdix perdix*.

Buteo buteo (Buzzard)
Palearctic. Up to 16 races described. Four birds are housed together whilst two others are housed separately. One of the singles is a piebald individual that was shot at Rise Park, Skirlaugh, Yorkshire on 1st January 1924.

Buteo lagopus (Rough-legged Buzzard)
Northern Holarctic. Four or five races. Two individual birds are housed separately.

FALCONIDAE

Falco tinnunculus (Kestrel)
Palearctic, Oriental and Ethiopian. At least 11 races. The single case holds an adult male.

Falco vespertinus (Red-footed Falcon)
Central and eastern Palearctic. Monotypic. The single case holds an adult male.

Falco subbuteo (Hobby)
Palearctic and, marginally, Oriental. Two races. Two birds are housed together.

GALLIFORMES

TETRAONIDAE

Lagopus lagopus (Willow/Red Grouse)
Holarctic. At least 17 races. Two birds, a male and female of the race *scoticus*, are cased together and were taken at Loch Doon, Ayrshire, Scotland in 1917. Another case contains three birds, a male and two females of the race *scoticus*, along with two small puppies that are shown hunting the grouse. This is a typical Victorian collage!

Tetrao urogallus (Capercaillie)
Western and central Palearctic. Up to ten races described. Two birds, a
male and female, are housed together.

PHASIANIDAE

Galloperdix sp. (Old English Game Cock)
Oriental. The single case contains a bird mounted with its spurs and
sparring gloves and bears the following inscription " *This bird, which
wears his original fighting spurs, won many a "main" in the middle of the XIXth
Century. The small sparring gloves were used for practising the bird with"*.

Syrmaticus reevesii (Reeve's Pheasant)
Endemic to northern and central China (introduced elsewhere).
Monotypic. The single case holds an adult male.

Phasianus colchicus (Pheasant)
Central and eastern Palearctic and Oriental (introduced to North
America and Europe, including Great Britain). Up to 34 races described.
A single case contains a very pale male (the paleness not being caused
by fading). Another case holds a male that was shot at Thryanby Hall
Wood (county untraced) on 23rd November 1923; the bird is white
throughout (save for a few dark flecks). Four males are housed together,
one of which is showing albinistic tendencies.

GRUIFORMES

GRUIDAE

Grus grus (Crane)
Palearctic. Two races.The single case holds an adult.

RALLIDAE

Rallus aquaticus (Water Rail)
Palearctic. Four races. A single case contains one bird. Another case holds
a single bird that was sexed as female upon dissection and, although the
collecting date is given as November 1915, no locality is indicated.

Porzana pusilla (Baillon's Crake)
Palearctic, Ethiopian and Australasian. Six or seven races. The single
case holds one bird.

Porphyrio porphyrio (Purple Gallinule)
Western Palearctic (patchily), Oriental, Ethiopian and Australasian. At least 16 races. The single case contains an adult that cannot be attributed to any particular race.

OTIDAE

Tetrax tetrax (Little Bustard)
Southwestern and central Palearctic. Probably monotypic, although some authors recognize two races. Two males and four females are housed together and are vaguely labelled as 'Old Yorkshire specimens'.

Otis tarda (Great Bustard)
Palearctic (patchily). Two, although some authors recognize three races. Three specimens are housed separately; a female, a male in breeding plumage and a non-breeding plumaged male. Although the female is labelled as "Rufforth, York, February 1861" it is known that it was shot on Rufforth Moor, the exact date being the 22nd of the month. The breeding plumaged male was taken at Staxton Wold, Yorkshire on an unrecorded date whilst the non-breeding plumaged male is devoid of data.

Chlamydotis undulata (Houbara Bustard)
Southwestern and central Palearctic and, marginally, Oriental. Three races. The single case holds an adult that was shot at Kirton Lindsey, Lincolnshire on 7th October 1847. This was the first recorded occurrence of this species in the British Isles, there being only four sightings since that time.

CHARADRIIFORMES

BURHINIDAE

Burhinus oedicnemus (Stone-curlew)
Southwestern Palearctic and Oriental. Up to eight races described. The single case contains one bird.

95

CHARADRIIDAE

Pluvialis apricaria (Golden Plover)
Northwestern Palearctic. Monotypic. Two races formerly described but now found to be clinal and not to warrant subspeciation. Two birds, both adults showing much summer plumage, are housed together.

Pluvialis squatarola (Grey Plover)
Northern Palearctic. Monotypic. Three birds, two in near full summer plumage the other in winter plumage, are housed together. The date on the back of the case is given as 30th August 1884 but no locality data is appended.

Charadrius morinellus (Dotterel)
Palearctic (mainly northern). Monotypic. Two cases, one with three birds the other with four.

SCOLOPACIDAE

Numenius arquata (Curlew)
Palearctic. Two races. Two birds are housed together.

Tringa totanus (Redshank)
Palearctic. Up to six races described. The single case contains two birds that were taken at Scarborough, Yorkshire on 4th September 1898.

Tringa flavipes (Lesser Yellowlegs)
Northern Nearctic. Monotypic. Two immature birds are housed together and were taken at Cleethorpes, Lincolnshire in 1889.

Tringa ochropus (Green Sandpiper)
Palearctic. Monotypic. The single case holds one bird.

Tringa glareola (Wood Sandpiper)
Palearctic. Monotypic. The single case contains one bird.

Phalaropus lobatus (Red-necked Phalarope)
Northern Holarctic. Monotypic. A single case contains a summer plumaged adult.

Phalaropus fulicarius (Grey Phalarope)
Northern Holarctic. Monotypic. Three birds, two moulting from juvenile to winter plumage and an adult that retains most of its summer plumage, are housed together. The immatures were acquired at

Easington, Yorkshire on an unrecorded date and the adult was taken at Scalby Ness, Scarborough, Yorkshire on 10th October 1896.

Gallinago media (Great Snipe)
Northwestern Palearctic. Monotypic. The single case contains four birds. One of the birds was acquired at Bolton Park, Yorkshire whilst the other three have no more precise data than being taken in Yorkshire. No dates are appended to any individual.

Gallinago gallinago (Snipe)
Holarctic, Ethiopian and Neotropical. Three races. Two birds are housed together.

Lymnocryptes minima (Jack Snipe)
Northern Palearctic. Monotypic. Two birds are housed together.

Calidris alba (Sanderling)
Holarctic. Monotypic. Four birds, three showing signs of summer plumage the other in winter plumage, are housed together. Two of the birds originate from Spurn, Yorkshire in May 1888 but the other two carry no data.

Calidris temminckii (Temminck's Stint)
Northern Palearctic. Monotypic. The single case holds one bird.

Calidris melanotos (Pectoral Sandpiper)
Northern Nearctic and northeastern Palearctic. Monotypic. The single case contains two birds that were taken at Charlestown, Firth of Forth, Scotland on an unrecorded adate.

Calidris alpina (Dunlin)
Holarctic. Six races. Five birds are housed together.

Calidris ferruginea (Curlew Sandpiper)
Northeastern Palearctic. Monotypic. Two winter plumaged birds are housed together. Although it is not possible to measure these birds, they are obviously at the two extremes of the species' bill length. Housed together they make a striking contrast and almost appear as different species.

Philomacus pugnax (Ruff)
Northern Palearctic. Monotypic. 17 birds,16 males and a female, are housed together. 14 of the male are in full summer plumage and show

the very variable colouration that this species exhibits whilst in this plumage, no two specimens being alike.

LARIDAE

Larus argentatus (Herring Gull)
Holarctic. At least five races. The single case holds an adult.

Larus ridibundus (Black-headed Gull)
Palearctic. Monotypic. The single case contains a first-winter bird.

Larus minutus (Little Gull)
Palearctic and, recently, Nearctic. Monotypic. A single case holds two juveniles. Another case contains three birds; two adults from Rudston, Yorkshire and an immature from Scarborough, Yorkshire (dates are not appended to any bird).

Xema sabini (Sabine's Gull)
Northern Holarctic. Monotypic. The single case contains three birds, which represent a juvenile, an adult in near full summer plumage and an adult in winter plumage.The winter plumaged bird was taken at Flamborough, Yorkshire on 5th September 1903 but the provenance of the others is not indicated.

STERNIDAE

Chlidonias nigra (Black Tern)
Western and central Palearctic and Nearctic. Two races. The single case holds an immature bird.

Sterna albifrons (Little Tern)
Cosmopolitan, though patchy in South America. Seven or eight races, although some American races are now considered a separate species (*Sterna antillarum*). Six birds, housed together, represent two adults and four small chicks.

ALCIDAE

Alle alle (Little Auk)
Northern Holarctic. Two races. The single case holds four birds that were shot on Filey Brigg, Yorkshire during the severe winter of 1894 - 95.

Alca torda (Razorbill)
North Atlantic. Two or three races. A single case holds a non-breeding plumaged bird.

Uria aalge (Guillemot)
Holarctic. Up to seven races described. Two summer plumaged adults of the unbridled form and two small chicks are housed together.

Cepphus grylle (Black Guillemot)
Northern Holarctic. Up to six races described. A single case holds a winter plumaged adult and an immature. Another case contains two breeding plumaged adults and an immature.

Fratercula arctica (Puffin)
North Atlantic. Three races. Three cases, one holding two birds the others containing singles.

COLUMBIFORMES

PTEROCLIDIDAE

Syrrhaptes paradoxus (Pallas's Sandgrouse)
Central Palearctic. Monotypic. Four cases, each with two birds, were acquired at the following localities; a male shot at Ellbin, near Beverley, Yorkshire on 19th May 1888 is housed with a female that was shot at Market Weighton, Yorkshire on 12th June 1888; a male and female shot at Bridlington, Yorkshire in June 1888 are housed together; a male and female from Spurn, Yorkshire on an unrecorded date are housed together, as are a male and female from the same locality that were taken in 1888. These latter birds are of the dark variety, being suffused with smokey-grey. This species is an extremely rare vagrant to our shores and the above birds were acquired during the great invasion of that year. It has been estimated that nearly 400 birds occurred in Yorkshire during this massive invasion.

COLUMBIDAE

Ectopistes migratorius (Passenger Pigeon)
Formerly the whole of North America, but now extinct. The last known bird, a female named Martha, died in Cincinnati Zoo on 1st September 1914; no wild birds had been reported for 12 years prior to this event. The single case holds an adult male.

STRIGIFORMES

TYTONIDAE

Tyto alba (Barn Owl)
Cosmopolitan (save for central and eastern Palearctic). At least 36 races.
The single case contains one bird.

STRIGIDAE

Otus scops (Scops Owl)
Palearctic and, marginally, Oriental. Up to 21 races described. Two birds,
housed together, are labelled "Scops Owl. Yorkshire. Strickland coll.". It
is known that Arthur Strickland took possession of a Scops Owl shot
near Driffield, Yorkshire in about 1839, but whether this bird is housed
in this case is a matter of conjecture.

Bubo bubo (Eagle Owl)
Palearctic, Oriental and, marginally, Ethiopian. At least 21 races. The
single case holds one bird.

Strix aluco (Tawny Owl)
Palearctic and northern Oriental. 12 or 13 races. The single case contains
a rufous phase bird.

Asio flammeus (Short-eared Owl)
Holarctic and Neotropical. Nine or ten races. The single case holds one
bird.

APODIFORMES

APODIDAE

Apus melba (Alpine Swift)
Southern Palearctic, Oriental and Ethiopian. At last ten races. The single
case contains a bird that had been picked up in an exhausted state at
Kirkburton, near Huddersfield, Yorkshire on 2nd June 1881. The bird
was sexed as a female upon dissection.

Apus apus (Swift)
Palearctic. Two races. Two birds, an adult and an immature, are housed
together.

CORACIIFORMES

ALCEDINIDAE

Alcedo atthis (Kingfisher)
Palearctic, Oriental and northern Australasian. At least eight races. The single case holds a bird that is displayed diving into a pool.

PICIFORMES

PICIDAE

Picus viridis (Green Woodpecker)
Western Palearctic. Up to 11 races described. Two cases hold individual birds that represent both sexes.

PASSERIFORMES

ALAUDIDAE

Lullula arborea (Woodlark)
Western Palearctic. Two races. Two birds are housed together.

Alauda arvensis (Skylark)
Palearctic (introduced to southwestern Australia and New Zealand). At least 16 races. A single case holds two birds. Another case contains a bird that is white throughout which was shot at Riccall, Yorkshire about 1870.

HIRUNDINIDAE

Riparia riparia (Sand Martin)
Holarctic and Oriental. At least five races. Three birds are housed together.

Hirundo rustica (Swallow)
Holarctic and northern Oriental. Eight races. The single case contains two adults.

Delichon urbica (House Martin)
Palearctic and northern Oriental. Five races. Two adults are housed together.

MOTACILLIDAE

Motacilla cinerea (Grey Wagtail)
Palearctic. Five or six races. A male and female-type are housed together.

Anthus pratensis (Meadow Pipit)
Western Palearctic. Two races. Two birds are housed together.

TURDIDAE

Erithacus rubecula (Robin)
Western Palearctic. Eight races. Three birds, representing a juvenile and two full grown individuals, are housed together.

Luscinia megarhynchos (Nightingale)
Southwestern Palearctic. Three races. The single case contains a bird which is erroneously labelled 'Great Reed Warbler *Acrocephalus arundinaceus* (Linn.)'.

Luscinia svecica (Bluethroat)
Palearctic and extreme northwestern Nearctic. Seven races. The single case contains a male. Although no collecting date is indicated, the bird is either moulting into, or out of, summer plumage. The throatal area is not fully coloured, thereby making subspecific determination impossible (even though the bird is labelled as being of the race *cyanecula*).

Phoenicurus ochruros (Black Redstart)
Western and central Palearctic and, marginally, Oriental. Up to seven races described. A male and two female-types are housed together.

Oenanthe oenanthe (Wheatear)
Palearctic and northern Nearctic. Up to seven races described. A male and female are housed together.

Zoothera dauma (White's Thrush)
Central and eastern Palearctic and Oriental. Up to 14 races described. The single case contains two birds.

102

EMBERIZIDAE

Emberiza rustica (Rustic Bunting)
Northern Palearctic. Two races. The single case contains a bird whose sex is impossible to determine.

Emberiza schoeniclus (Reed Bunting)
Palearctic (save for the northeast). Up to 15 races described. A male and female are housed together.

Plectrophenax nivalis (Snow Bunting)
Holarctic. Four races. The single case holds nine birds that depict winter plumage of both sexes.

FRINGILLIDAE

Carduelis carduelis (Goldfinch)
Western and central Palearctic and, marginally, Oriental (introduced elsewhere). Up to 19 races described. The single case contains an adult bird.

ORIOLIDAE

Oriolus oriolus (Golden Oriole)
Western and central Palearctic and northern Oriental. Two races. The single case contains an adult male.

CORVIDAE

Garrulus glandarius (Jay)
Palearctic and northern Oriental. Up to 36 races described. Two birds are housed separately.

Pica pica (Magpie)
Holarctic, Oriental and Ethiopian (Saudi Arabia only). 13 races. Two cases hold individual birds.

Pyrrhocorax pyrrhocorax (Chough)
Southern Palearctic and northeastern Ethiopian. Seven or eight races. The single case contains an adult.

Corvus monedula (Jackdaw)
Western and central Palearctic. Four races. The single case holds one bird.

Corvus frugilegus (Rook)
Palearctic. Two race. The single case contains an adult bird.

Corvus corone (Carrion Crow)
Palearctic and, marginally, Oriental. Six races. Three birds; a single of the race *cornix* and two of the nominate race which are housed together.

MULTI-SPECIES CASED MOUNTS

The following cases house more than one species of bird:

(1). A single *Tachybaptus ruficollis* (Little Grebe) and a *Podiceps auritus* (Slavonian Grebe), both birds are in winter plumage.

(2). Two female *Mergus merganser* (Goosander), two summer plumaged *Limosa lapponica* (Bar-tailed Godwit) and five breeding plumaged males and a female *Philomacus pugnax* (Ruff).

(3). A female *Tetrao urogallus* (Capercaillie) and a hybrid male *Tetrao tetrix* X *Tetrao urogallus* (Black Grouse X Capercaillie). Neither bird carries collecting data but the hen is said to be assumimg cock's plumage and the hybrid is said to be a hen. Both these statements are nonsense!

(4). A male *Phasianus colchicus* (Pheasant) and a very unusual partner - a female *Regulus regulus* (Goldcrest).

(5). A male and female *Syrmaticus reevesii* (Reeve's Pheasant) and a mainly summer plumaged female *Lagopus mutus* (Ptarmigan).

(6). A case, labelled "Ringed Plover *Charadrius hiaticula* Linnaeus", contains an adult of this species and a male *Charadrius alexandrinus* (Kentish Plover).

(7). An adult and an immature *Charadrius hiaticula* (Ringed Plover), a summer plumage *Arenaria interpres* (Turnstone), a winter plumaged *Calidris alpina* (Dunlin) and a mainly winter plumaged *Calidris alba* (Sanderling).

(8). A single *Arenaria interpres* (Turnstone) and a *Calidris canutus* (Knot), both birds are mainly in breeding plumage.

(9). An adult female and immature female *Falco peregrinus* (Peregrine) and a winter plumaged male *Lagopus mutus* (Ptarmigan).

(10). An immature *Stercorarius parasiticus* (Arctic Skua), two immature *Larus argentatus* (Herring Gull), two immature *Larus canus* (Common Gull) and two immature *Sterna hirundo* (Common Tern).

(11). A winter plumaged adult *Rissa tridactyla* (Kittiwake) and a winter plumaged *Alle alle* (Little Auk).

(12). An adult *Hirundo rustica* (Swallow) and a juvenile *Riparia riparia* (Sand Martin).

(13). A single *Erithacus rubecula* (Robin), *Prunella modularis* (Dunnock), *Parus caeruleus* (Blue Tit), male *Passer domesticus* (House Sparrow) and a male *Carduelis chloris* (Greenfinch).

(14). A single *Passer montanus* (Tree Sparrow), female *Emberiza schoeniclus* (Reed Bunting), *Carduelis carduelis* (Goldfinch), male *Pyrrhula pyrrhula* (Bullfinch) and a male and female *Carduelis chloris* (Greenfinch).

(15). A single male *Emberiza citrinella* (Yellowhammer), male *Fringilla coelebs* (Chaffinch) and a *Carduelis carduelis* (Goldfinch).

(16). A single *Perdix perdix* (Grey Partridge), a male and female *Passer domesticus* (House Sparrow), *Sturnus vulgaris* (Starling) and a duckling *Somateria mollissima* (Eider). A *Mustela erminea* (Stoat) is displayed attacking the *Perdix perdix*.

(17). Two *Strigops habroptilus* (Kakapo), which are said to be male and female, *Apteryx australis* (Brown Kiwi) and an *Aptertx owenii* / *Apteryx haastii* (Little Spotted Kiwi / Great Spotted Kiwi). These New Zealand endemics were shot by the Rev. Charles M. Meysey Thompson in 1880. The identification of *A. owenii* and *A. haastii*, even in the hand, is fraught with difficulty. Plumage details and biometrics, especially of immature birds, offer little hope of specific identification. *The Handbook of Australasian, New Zealand & Antarctic Birds* (1990) indicates that the best ways to identify these species are by calls and the species specific parasites which they harbour - this specimen has been dead for over a century and can, unfortunately, offer neither!

Two bamboo framed cases contain the following Australasian and Oriental species, several of which are endemic to Australia:

(1). A single adult *Chrysococcyx lucidus* (Shining Bronze-cuckoo), a female *Halcyon macleayii* (Forest Kingfisher), a male *Petroica rosea* (Rose Robin), an adult *Eopsaltria australis* (Eastern Yellow Robin), two male *Pachycephala pectoralis* (Golden Whistler), an adult *Cormobates leucophaea* (White-throated Treecreeper), a male *Dicaeum hirundinaceum* (Mistletoebird), a male and female/immature *Pardalotus punctatus* (Spotted Pardalote), an adult *Pardalotus striatus* (Striated Pardalote) and a *Erythrura gouldiae* (Gouldian Finch). The *Pardalotus striatus* shows a near uniform black crown and may belong to the race *melanocephalus*.

(2). A single *Turnix varia* (Painted Button-quail), an adult *Chalcophaps indica* (Emerald Dove), an adult *Cuculus pyrrhophanus* (Fan-tailed Cuckoo), a male and female *Eudynamys scolopacea* (Koel), *Eurystomus orientalis* (Dollarbird), an adult *Coracina lineata* (Yellow-eyed Cuckoo-shrike), a male *Sericulus chrysocephalus* (Regent Bowerbird) and an adult *Cracticus torquatus* (Grey Butcherbird).

UNCASED MOUNTS

The Museum possesses a total of 403 uncased mounts that represent 179 species.

STRUTHIONIFORMES

STRUTHIONIDAE

Struthio camelus (Ostrich)
Ethiopian. Five or six races. A single downy chick standing 230mm high.

APTERYGIFORMES

APTERYGIDAE

Apteryx australis (Brown Kiwi)
Endemic to New Zealand. Three races. Two adults.

Apteryx haastii (Great Spotted Kiwi)
Endemic to New Zealand. Monotypic. The single adult was collected in New Zealand.

Apteryx owenii / Apteryx haastii (Lesser Spotted / Great Spotted Kiwi)
Both species are endemic to New Zealand and monotypic. The single bird is labelled *"Apteryx oweni,* Gould. (Young). New Zealand. E. Coll. R.W.Linfoot". The identification of *A. owenii* from *A. haastii,* even in the hand, is fraught with difficulty. Plumage details and biometrics, especially of immature birds, offer little hope of specific identification. *The Handbook of Australasian, New Zealand & Antarctic Birds* (1990) indicates that the best ways to identify these species are by call and the species specific parasites which they harbour - this specimen has probably been dead for over a century and can, unfortunately, offer neither!

GAVIIFORMES

GAVIIDAE

Gavia stellata (Red-throated Diver)
Northern Holarctic. Monotypic. Four birds; one in summer plumage, two in winter plumage and one in transitional plumage. One of the birds,

sexed as male upon dissection, was found in September 1986 but bears no locality data.

Gavia arctica (Black-throated Diver)
Northern Holarctic. Three races, although some authors now regard *pacifica* as a separate species. Two adults in full breeding plumage.

Gavia immer (Great Northern Diver)
Northern Holarctic and, marginally, Palearctic. Monotypic. Three adults in summer plumage. One of the birds is either beginning to attain winter plumage on the cheeks and throat or has still to acquire its full plumage.

PODICIPEDIFORMES

PODICIPEDIDAE

Podiceps grisegena (Red-necked Grebe)
Holarctic. Two Races. Five birds. Four birds, a juvenile, one in winter plumage and two in summer plumage had originally been housed together. The other, a bird in winter plumage, had originally been cased with a similarly plumaged bird but the whereabouts of this bird is unknown.

Podiceps cristatus (Great Crested Grebe)
Palearctic, Oriental, Ethiopian (patchily) and Australasian. Three races. A juvenile and an adult in summer plumage had originally been housed together.

Podiceps auritus (Slavonian Grebe)
Holarctic. Monotypic. The two birds, one in summer plumage the other in winter plumage, had originally been cased together.

PROCELLARIIFORMES

PROCELLARIIDAE

Fulmarus glacialis (Fulmar)
Holarctic. At least two races. The single light phase bird is displayed in the gliding position.

Bulweria bulwerii (Bulwer's Petrel)
Islands in the Atlantic and Pacific Oceans. Monotypic. A single bird.

Puffinus griseus (Sooty Shearwater)
Neotropical (southern South America) and Australasian (New Zealand, southeastern Australia and Tasmania). Monotypic. The two birds had originally been housed together.

Puffinus puffinus (Manx Shearwater)
Western Palearctic (mainly Great Britain), Nearctic (patchily) and Australasian (New Zealand only). Six races, although some authors now regard the five in the Pacific as separate species. Two specimens of the nominate race.

PELECANIFORMES

SULIDAE

Morus bassanus (Gannet)
Seaboards on both sides of the North Atlantic. Monotypic. Three adults and a single juvenile.

PHALACROCORACIDAE

Phalacrocorax carbo (Cormorant)
Almost cosmopolitan, though not found in the Neotropics. At least five races. Two birds; a sub-adult and a breeding plumaged adult of the nominate race, that had previously been housed in a case.

Phalacrocorax aristotelis (Shag)
Western Palearctic. Three races. Two birds; an adult in breeding plumage and an immature.

CICONIIFORMES

ARDEIDAE

Ardea purpurea (Purple Heron)
Palearctic, Oriental and Ethiopian. Four races. The summer plumaged adult and an immature had originally been cased together.

Ardea cinerea (Grey Heron)
Palearctic, Oriental and Ethiopian. At least four races. Seven birds; four adults in summer plumage and three immatures. One of the adults is

displayed in flight and another had originally been cased. Two of the immatures had been housed together.

ANSERIFORMES

ANATIDAE

Cygnus olor (Mute Swan)
Palearctic (introduced in Nearctic). Monotypic. The single adult had originally been cased.

Cygnus columbianus (Bewick's Swan)
Northern Holarctic. Two races. Two adults of the race *bewickii* ; one of which is displayed in flight.

Cygnus cygnus (Whooper Swan)
Palearctic. Monotypic. The single adult had originally been cased.

Anser brachyrhynchus (Pink-footed Goose)
Greenland, eastern Iceland and Spitsbergen only. Monotypic. The single bird had previously been housed in a case.

Anser albifrons (White-fronted Goose)
Northern Holarctic. Five races. The single adult, of indecipherable race, had originally been cased.

Anser anser (Greylag Goose)
Palearctic. Two races. The single bird, that is displayed with an extended right wing, had originally been housed in a case. The bird's race cannot be ascertained.

Branta leucopsis (Barnacle Goose)
Eastern Greenland, Spitsbergen and Novaya Zemlya only. Monotypic. The single bird had previously been cased.

Branta bernicla (Brent Goose)
Northern Holarctic. At least three races. The single adult of the race *hrota* had originally been cased.

Chloephaga picta (Magellan Goose)
Endemic to southern South America and the Falkland Islands. Two distinct races which may be best referred to as species. The single bird, a

first-year male, belongs to the race *leucoptera* (often called Greater or Falkland Island Magellan Goose).

Tadorna tadorna (Shelduck)
Northwestern and southern Palearctic. Monotypic. The two birds had previously been housed together.

Anas penelope (Wigeon)
Northern Palearctic. Monotypic. The two males and a female had originally been cased together.

Anas crecca (Teal)
Holarctic. Two, perhaps 3 races. A male (of the nominate race) and a female had previously been housed together.

Anas platyrhynchos (Mallard)
Holarctic. Seven races. Two males and a female. One of the males is white throughout and no doubt stems from domestic stock. For some unknown reason the other male has "Teal" written on the bottom of the base.

Anas acuta (Pintail)
Holarctic with isolated populations in the southern Indian Ocean. Three races. The two males and a female had originally been cased together.

Anas platyrhynchos (Mallard) X *Anas acuta* (Pintail)
A male bird that had originally been cased is labelled " *HYBRID between Mallard and Pintail. Locality - Lowthorpe. Presented by W.H.St. Quintin. 1916"*. This Yorkshire bird has the curved central tail feathers indicative of male Mallard and certain features that fit Pintail. Seen in the wild, however, this individual would simply appear as a 'man-made park duck'.

Anas clypeata (Shoveler)
Holarctic. Monotypic. The single male is not in full summer plumage.

Netta rufina (Red-crested Pochard)
Western (patchily) and central Palearctic. Monotypic. A single female-type.

Aythya ferina (Pochard)
Palearctic. Monotypic. Two males and two females.

Aythya fuligula (Tufted Duck)
Palearctic. Monotypic. The male and female had previously been housed together.

Aythya marila (Scaup)
Northern Holarctic. At least two races. The male and female had originally been cased together. The male was taken at Scarborough, Yorkshire in 1899 and the female in Wensleydale, Yorkshire on an unrecorded date.

Somateria mollissima (Eider)
Northern Holarctic. Six races. Two males and two females. The two males and one of the females had previously been housed together and the other female is displayed in flight.

Melanitta nigra (Common Scoter)
Northern Palearctic and northern Nearctic (patchily). Two races. A male of the nominate race had originally been cased.

Melanitta fusca (Velvet Scoter)
Northern Holarctic. At least three races. The single male belongs to the nominate race.

Mergus cucullatus (Hooded Merganser)
Nearctic. Monotypic. The single male, from Ferriby Sluice, Yorkshire in 1889, had previously been cased. This specimen is of interest as the species is but an accidental visitor to this side of the Atlantic. As the chances of this bird being an escape from captivity or even fraudulently labelled are high, no attempt will be made to submit details to the relevant authorities claiming this bird as a genuine vagrant.

Mergus albellus (Smew)
Palearctic. Monotypic. The adult male and female-type birds had originally been housed together.

Mergus serrator (Red-breasted Merganser)
Northern Holarctic. Monotypic. The single male had originally been cased.

Mergus merganser (Goosander)
Holarctic. Three races. Two males and two females. A male and female had previously been housed together and the other male is displayed in flight.

113

FALCONIFORMES

PANDIONIDAE

Pandion haliaetus (Osprey)
Cosmopolitan, though patchy in Ethiopian and Neotropics. Five or six described races. A single adult.

ACCIPITRIDAE

Pernis apivorus (Honey Buzzard)
Western Palearctic. Monotypic. A single bird.

Milvus milvus (Red Kite)
Western Palearctic. Two races. An adult and a first-winter. The base on which the adult is mounted simply indicates that it was taken along the coast of Spain.

Haliaeetus albicilla (White-tailed Sea Eagle)
Palearctic and Nearctic (southwest Greenland). Monotypic. An adult and three immatures. Two of the immatures had originally been housed together.

Circus aeruginosus (Marsh Harrier)
Palearctic, Ethiopian (Madagascar only) and Australasian. Nine races. One male and three female-types. The male and one of the female-types had originally been housed together.

Circus cyaneus (Hen Harrier)
Holarctic. Four races. Three female-types, two of which had previously been cased together and were taken at Pocklington, Yorkshire in about 1878. The other is displayed with its next meal, a Lapwing *Vanellus vanellus*.

Accipiter gentilis (Goshawk)
Holarctic. Between seven and nine races described, although some authors regard the North American races as a separate species (*Accipiter atricapillus*). An adult and an immature had originally been housed together.

Accipiter nisus (Sparrowhawk)
Palearctic. Six races. Two adult males, two first-year males and two first-year females. One of the females is displayed attacking two Starlings

Sturnus vulgaris. One of the adult males is displayed on a gamekeeper's gibbet and is wearing a British Trust for Ornithology ring (number DA52901). The bird had originally been ringed as an adult at Filey, Yorkshire on 22nd May 1984 and had killed itself by flying into a window at Reighton, Yorkshire on 11th April 1986 (some 5 Kms. south). The bird is depicted as a gamekeeper's victim simply to gain the desired effect!

Buteo lagopus (Rough-legged Buzzard)
Northern Holarctic. Four or five races. Two birds. One had previously been housed with another but the whereabouts of this bird is unknown.

Aquila chrysaetos (Golden Eagle)
Holarctic, marginally Oriental and Ethiopian. The single adult had originally been cased.

FALCONIDAE

Falco naumanni (Lesser Kestrel)
Southern Palearctic. Monotypic. A second-year male. This bird, shot at Green Hammerton, Yorkshire about the middle of November 1867, ensured the species a place on the British list. The claws of this individual are somewhat short, a sign that it may have been kept in confinement, although it is not always possible to be certain.

Falco tinnunculus (Kestrel)
Palearctic, Oriental and Ethiopian. At least 11 races. Two adult males and two first-year males. One of the adults is displayed attacking a Wood Mouse *Apodemus sylvaticus* and a first-year is depicted in the hovering position.

Falco columbarius (Merlin)
Holarctic. 11 races. A single bird.

Falco subbuteo (Hobby)
Palearctic and, marginally, Oriental. Three birds, of which two had previously been housed together.

Falco rusticolus (Gyrfalcon)
Holarctic. Monotypic. Three birds; an adult light morph, an immature light morph and one of indeterminable age and morph caused by fading. The immature was originally housed with two others but the whereabouts of these birds is unknown.

Falco cherrug (Saker)
Central and western Palearctic. Two races, although some authors recognize four. The single bird has recently been donated (December 1993) by the Al-Areen Wildlife Park, Bahrain. As the bird had originally been held in captivity, its geographical origin remains uncertain.

Falco peregrinus (Peregrine)
Cosmopolitan. At least 15 races. Three birds; an immature female, a second-year female and an adult male. The second-year female has recently been donated (December 1993) by the Al-Areen Wildlife Park, Bahrain. As the bird had originally been held in captivity, its geographical origin remains uncertain.

GALLIFORMES

TETRAONIDAE

Lagopus lagopus (Willow/Red Grouse)
Holarctic. At least 17 races. The male and four females belong to the race *scoticus*.

Lagopus mutus (Ptarmigan)
Holarctic. At least 23 races. Six birds (two males and two females in summer plumage and two females in winter plumage). The summer plumaged birds had originally been housed together, as had the winter plumaged birds.

Tetrao tetrix (Black Grouse)
Palearctic. Up to eight races described. Two males in breeding plumage and three females.

Tetrao urogallus (Capercaillie)
Western and central Palearctic. Up to ten races described. The adult male and female had previously been housed together.

PHASIANIDAE

Alectoris rufa (Red-legged Partridge)
Southwestern Palearctic (introduced to Great Britain). Up to five races described. A single bird.

Phasianus colchicus (Pheasant)
Central and eastern Palearctic and Oriental (introduced to North America and Europe, including Great Britain). Up to 34 races described. Six males and two females. One of the males is white throughout save for a few dark flecks. A single male had previously been cased, as had a male labelled *"CHINESE PHEASANT* Phasianus colchicus. *Old Hen assuming male plumage. Locality - THORNTON - LE - DALE. Presented by R. Hill. 1910"*. On the surface, this Yorkshire bird appears simply to be a cock Pheasant!

Chrysolophus pictus (Golden Pheasant)
Southeastern Palearctic (indigenous populations only in mountains of central China). Monotypic. A single male.

Pavo cristatus (Peafowl)
Oriental. Monotypic. The single cock, which is displayed with its tail fanned, originates from a semi-feral population in the Museum Gardens, York, Yorkshire.

GRUIFORMES

RALLIDAE

Rallus aquaticus (Water Rail)
Palearctic. Four races. Two birds.

Crex crex (Corncrake)
Palearctic. Monotypic. Three birds. One had originally been cased and the other two had previously been housed together

Porzana porzana (Spotted Crake)
Western and central Palearctic. Monotypic. Three birds had previously been housed together.

Gallinula chloropus (Moorhen)
Almost cosmopolitan (save for Australasian). Up to 12 races described. Five birds. One bird is displayed with its neck outstretched and its wings open ready for takeoff. Two birds, which may have been taken at Heworth, York, Yorkshire in December 1890, had previously been housed together.

Fulica atra (Coot)
Palearctic, Oriental and Ethiopian. Four races. The two birds had originally been cased together.

OTIDAE

Otis tarda (Great Bustard)
Palearctic (patchily). Two, although some authors recognize three races.
The single male is in non-breeding plumage.

CHARADRIIFORMES

HAEMATOPODIDAE

Haematopus ostralegus (Oystercatcher)
Palearctic. Three races. The single adult is in summer plumage.

RECURVIROSTRIDAE

Recurvirostra avosetta (Avocet)
Palearctic and Ethiopian. The single bird, which had originally been
cased, was shot at Scarborough, Yorkshire in about 1865.

BURHINIDAE

Burhinus oedicnemus (Stone-curlew)
Southwestern Palearctic and Oriental. Up to eight races described. The
two adults and two small chicks from Norfolk on an unrecorded date
were originally housed together.

GLAREOLIDAE

Glareola pratincola (Collared Pratincole)
Southern Palearctic and Ethiopian (patchily). Up to five races described.
The single bird had previously been cased.

CHARADRIIDAE

Vanellus vanellus (Lapwing)
Palearctic. Monotypic. Three birds, two of which had originally been
housed together.

SCOLOPACIDAE

Limosa limosa (Black-tailed Godwit)
Palearctic. Three races. Five birds (two in summer plumage the others in winter plumage). The three winter plumaged birds and a summer plumaged bird had previously been housed together.

Limosa lapponica (Bar-tailed Godwit)
Northern Palearctic and northwestern Nearctic. Two races. Four birds (one in summer plumage the others in winter plumage). One of the summer plumaged birds is displayed with its wings half spread whilst two of the winter plumaged birds had originally been housed together.

Tringa erythropus (Spotted Redshank)
Northern Palearctic. Monotypic. A single winter plumaged bird.

Tringa ochropus (Green Sandpiper)
Palearctic. Monotypic. Two birds. One of the birds had previously been housed with two Common Sandpiper *Actitis hypoleucos* and another Green Sandpiper. The whereabouts of one of the Common Sandpiper and the other Green Sandpiper is unknown (see *Actitis hypoleucos* below).

Actitis hypoleucos (Common Sandpiper)
Palearctic. Monotypic. Two birds. One of the birds had previously been housed with two Green Sandpiper *Tringa ochropus* and another Common Sandpiper. The whereabouts of one of the Green Sandpiper and the other Common Sandpiper is unknown (see *Tringa ochropus* above).

Arenaria interpres (Turnstone)
Northern Holarctic. Two races. Five birds, which are mainly in summer plumage but much faded, had originally been housed together.

Scolopax rusticola (Woodcock)
Palearctic. Monotypic. Four birds. One was taken at Spurn, Yorkshire on an unrecorded date and another had previously been in a case.

Calidris canutus (Knot)
Holarctic. Four races. The single juvenile is displayed in flight.

Calidris maritima (Purple Sandpiper)
Northern Holarctic. Monotypic. The three birds, which exhibit varying amounts of summer plumage, had originally been cased together.

Calidris alpina (Dunlin)
Holarctic. Six races. The four birds, one of which is in summer plumage, were collected at Easington, Yorkshire (no date appended) and had originally been housed together.

Philomacus pugnax (Ruff)
Northern Palearctic. Monotypic. Three breeding plumaged males and a female. Two of the males and the female had previously been cased together.

STERCORARIIDAE

Stercorarius pomarinus (Pomarine Skua)
Northern Holarctic. Monotypic. The four birds, representing two pale phase adults and two immatures, had originally been housed together. The adults originate from Spurn, Yorkshire in May 1887 and Lapland (no country indicated) in June 1888 whilst the immatures were taken at Bridlington, Yorkshire (no date stated) and Harome, Yorkshire in December 1895. Inland records of this species, as indicated by this latter bird, are an unusual occurrence.

Stercorarius parasiticus (Arctic Skua)
Northern Holarctic. Monotypic. The pale phase adult had originally been cased.

Stercorarius longicaudus (Long-tailed Skua)
Northern Holarctic. Two races. The adult and an immature had previously been cased together.

LARIDAE

Larus canus (Common Gull)
Palearctic and northwestern Nearctic. Four races. Two birds, a first-winter and a second-winter, had originally been housed together.

Larus fuscus (Lesser Black-backed Gull)
Northwestern Palearctic. Three races. The single adult of the race *graellsii* had previously been cased.

Larus marinus (Great Black-backed Gull)
Northern Atlantic. Monotypic. Five birds, representing first-winter (2), second-winter (1) and adult (2). The first-winter birds were collected at Spurn, Yorkshire in January 1896 and had originally been housed together. The adults were shot at Bridlington, Yorkshire on 24th January 1895 and 25th July 1915 and had originally been displayed in a single case.

Larus hyperboreus (Glaucous Gull)
Northern Holarctic. Three races. The adult and a sub-adult had originally been housed together.

Larus glaucoides (Iceland Gull)
Northern Nearctic (southern Baffin Island and Greenland only). Two races. The adult of the nominate race had originally been cased.

Larus ridibundus (Black-headed Gull)
Palearctic. Monotypic. Seven birds, representing pullus, first-winter, adult winter, adult summer and adult in partial summer plumage. The first-winter, adult winter and partial summer plumaged adult had previously been housed together. The summer plumaged adult is displayed in flight.

Rissa tridactyla (Kittiwake)
Northern Holarctic. Two races. Three birds. A summer plumaged adult and two second-winter birds had originally been cased together.

ALCIDAE

Alle alle (Little Auk)
Northern Holarctic. Two races. The three winter plumaged birds had originally been housed together.

Alca torda (Razorbill)
North Atlantic. Two or three races. Five birds (four in winter plumage the other in summer plumage). The summer plumaged bird and a winter plumaged bird had previously been cased together.

Pinguinus impennis (Great Auk)
Former breeding stations known to include Bird Rocks in the Gulf of St. Lawrence; Funk Island, off the New Foundland coast; Grimsey (no longer extant due to volcanic activity) and Eldney Islands near Iceland and St. Kilda, Hebrides, Scotland. The last birds are thought to have been

121

killed on 3rd June 1844 on the island of Eldney. The two adults had originally been in cases.

Uria aalge (Guillemot)
Holarctic. Up to seven races described. Three winter plumaged birds.
One of the birds, from Filey, Yorkshire in January 1895, had been used
by Archibolt Thorburn to illustrate Brunnich's Guillemot *Uria lomvia* in
Lord Lilford's book *Coloured Figures of the Birds of the British Islands* (1891
- 97). This error in identification was brought to light in a paper by
R.Wagstaffe, K.Williamson and R.H.Broughton in *The North Western
Naturalist* published in March 1946. The other two birds had previously
been housed together.

Cepphus grylle (Black Guillemot)
Northern Holarctic. Up to six races described. The two breeding
plumaged birds had previously been housed together.

COLUMBIFORMES

COLUMBIDAE

Columba livia (Rock Dove)
Originally southern Palearctic, Oriental and northern Ethiopian, but
now almost cosmopolitan due to feral populations. Up to 14 races
described. The two birds, which appear to be pure bred i.e. without the
interference of Man, were taken at Bridlington, Yorkshire in 1895. They
had originally been cased together.

Columba oenas (Stock Dove)
Western Palearctic. Two or three races. The three birds had previously
been housed together.

Columba palumbus (Woodpigeon)
Western Palearctic with isolated populations in central Palearctic. Five
or six races. The two adults had originally been housed together.

Streptopelia turtur (Turtle Dove)
Western and central Palearctic and, marginally, Ethiopian. Four races.
The two birds had previously been housed together.

Goura cristata (Blue-crowned Pigeon)
Endemic to northwestern New Guinea and a few outlying islands. Two
races. A single adult.

PSITTACIFORMES

LORIIDAE

Eos reticulata (Blue-streaked Lory)
Endemic to the Tanimbar Islands, Indonesia (introduced to the Kai Islands and Damar Island, Indonesia). Monotypic. The single adult is displayed with outstretched wings. Although no data is attached, it is known that the bird is from captive stock.

Trichoglossus haematodus (Rainbow Lory)
Australasian. 21 races. The single adult belongs to the nominate race. Although no data is attached, it is known that the bird is from captive stock.

Charmosyna papou (Papuan Lory)
Endemic to New Guinea. Four races. The single bird belongs to one of the three races which exhibit a melanic phase. Subspecific determination is clouded, however, as it is known that the bird is from captive stock and may, therefore, not be racially pure.

PSITTACIDAE

Eclectus roratus (Eclectus Parrot)
Australasian (New Guinea and surrounding islands and extreme northern Queensland, Australia). Ten races. The adult female is of one of the races which exhibit yellow undertail-coverts. Subspecific determination is clouded, however, as it known that the bird is from captive stock and may, therefore, not be racially pure.

Psittacus erithacus (African Grey Parrot)
Ethiopian (central Africa only). Three races. The single adult is displayed with an outstretched wing and scratching its head. Although no data is attached, it is known that the bird is from captive stock.

Amazona amazonica (Orange-winged Amazon)
Neotropical (northern South America only). Two races. The single bird is of the nominate race and is displayed with its wings half extended. Although no data is attached, it is known that the bird is from captive stock.

CUCULIFORMES

CUCULIDAE

Cuculus canorus (Cuckoo)
Palearctic and northern Oriental. Up to nine races described. The single adult and juvenile had originally been housed together.

STRIGIFORMES

TYTONIDAE

Tyto alba (Barn Owl)
Cosmopolitan (save for central and eastern Palearctic). At least 36 races. Five birds (four males and an unsexable bird). Two males had previously been housed together, another male is displayed with wings outstretched in an attacking position and the other male is displayed in flight carrying an *Apodemus sylvaticus* (Wood Mouse). The unsexable bird is displayed in flight, and featured in the Yorkshire Evening Press of 11th January 1989, having been presented to the Museum by Stockton Hall School.

STRIGIDAE

Bubo bubo (Eagle Owl)
Palearctic, Oriental and, marginally, northern Ethiopian. At least 21 races. Two birds.

Nyctea scandiaca (Snowy Owl)
Northern Holarctic. Monotypic. A single female and two recently fledged juveniles. The juveniles were taken at Stjordal, Norway in July 1891.

Strix aluco (Tawny Owl)
Palearctic and northern Oriental. 12 or 13 races. Four birds. A recently fledged downy chick that was collected at Scarborough, Yorkshire on an unrecorded date had originally been cased; a bird with a grey ground colour was taken at South Milford, Yorkshire on 23rd October 1985; a bird with a rufous ground colour is displayed in a pole trap and another bird with a rufous ground colour is displayed in flight with a male Chaffinch *Fringilla coelebs* as prey. The other bird has a rufous ground colour.

124

Asio otus (Long-eared Owl)
Holarctic and Ethiopian (patchily). Up to six races described. Four birds, of which one is displayed in flight.

Asio flammeus (Short-eared Owl)
Holarctic and Neotropical. Nine or ten races. A single bird.

CAPRIMULGIFORMES

CAPRIMULGIDAE

Caprimulgus europaeus (Nightjar)
Western and central Palearctic. Up to ten races described. A male and female along with two small chicks were originally housed together. The male is displayed in flight.

CORACIIFORMES

UPUPIDAE

Upupa epops (Hoopoe)
Palearctic, Oriental and Ethiopian. Ten races. The two birds had previously been housed together.

BUCEROTIDAE

Aceros undulatus (Wreathed Hornbill)
Oriental (Assam, Indochina, Malay Peninsula and Sunda Islands only). Two races. Two birds.

Buceros rhinoceros (Rhinoceros Hornbill)
Oriental (Malay Peninsula, Sunda Islands and Borneo only). Four races. The single immature is not sporting the fully developed casque indicative of the species.

Buceros bicornis (Great Indian Hornbill)
Oriental (eastern India to Thailand, south to Sumatra only). Two races. A single adult.

PICIFORMES

RAMPHASTIDAE

Rhamphastos toco (Toco Toucan)
Neotropical. Two races. A single bird.

PICIDAE

Jynx torquilla (Wryneck)
Palearctic. Up to ten races described. Three birds. One of the birds was sexed as a male upon dissection and another is displayed in flight having originally been cased.

Dendrocopus minor (Lesser Spotted Woodpecker)
Palearctic. Up to 19 races described. A single male and female. The male is displayed in flight.

Dendrocopus major (Great Spotted Woodpecker)
Palearctic and, marginally, Oriental. Up to 31 races described. Two males and two females. One of the females is displayed in flight.

Picus viridis (Green Woodpecker)
Western Palearctic. Up to 11 races described. Four birds, representing juvenile, male and female (2). Two birds, a female and the juvenile, had originally been cased together; the female is displayed with its wings half open, the juvenile in full flight.

PASSERIFORMES

EURYLAIMIDAE

Calyptomena viridis (Lesser Green Broadbill)
Oriental (Lower Burma and peninsula Thailand to Sumatra and Borneo only). Three races. A single bird.

CINCLIDAE

Cinclus cinclus (Dipper)
Palearctic. Up to 11 races described. A single bird.

126

TROGLODYTIDAE

Troglodytes troglodytes (Wren)
Holarctic and, marginally, Oriental. Up to 37 races described. A single bird.

PRUNELLIDAE

Prunella modularis (Dunnock)
Western Palearctic. Eight races. A single bird.

TURDIDAE

Erithacus rubecula (Robin)
Western Palearctic. Eight races. Six birds. Two adults, one of which is displayed with four nearly fledged young in a nest, had originally been housed in a domed case (now broken and discarded).

Phoenicurus phoenicurus (Redstart)
Western and central Palearctic. Two or three races. Three birds; an adult male, female and juvenile that had originally been housed together.

Saxicola rubetra (Whinchat)
Western Palearctic. Monotypic. Two males and a single in female-type plumage.

Saxicola torquata (Stonechat)
Palearctic, Ethiopian and, marginally, Oriental. 25 races. Two male and a female-type. The female-type is erroneously labelled as a Whinchat *Saxicola rubetra*.

Zoothera dauma (White's Thrush)
Central and eastern Palearctic and Oriental. Up to 14 races described. The single bird had originally been housed in a case labelled "Mistle Thrush". Is this a hitherto unrecorded British, or even Yorkshire, bird? Unfortunately the answer to this question is very unlikely to be forthcoming!

Turdus merula (Blackbird)
Western and southern Palearctic and, marginally, Oriental (introduced to southeastern Australia and New Zealand). 16 races. Two birds; an adult male and an adult female.

127

Turdus philomelos (Song Thrush)
Western and central Palearctic. Four races. The single bird is displayed sitting on a nest.

Turdus iliacus (Redwing)
Palearctic. Two races. A single bird.

SYLVIIDAE

Sylvia borin (Garden Warbler)
Western and central Palearctic. Three races. A single bird.

Sylvia atricapilla (Blackcap)
Western Palearctic. Five races. The single female was killed by a car at Haxby, Yorkshire on 14th April 1980.

Sylvia communis (Whitethroat)
Western and central Palearctic. A single bird.

Phylloscopus trochilus (Willow Warbler)
Palearctic. Three races. Three birds. One of the birds, despite having its legs painted black to make it appear as a Chiffchaff *Phylloscopus colybita*, has the wing formula indicative of *P. trochilus*.

Phylloscopus sibilatrix (Wood Warbler)
Western Palearctic. Monotypic. A single bird.

Regulus regulus (Goldcrest)
Palearctic. Up to 14 races described. A single male.

MUSCICAPIDAE

Ficedula hypoleuca (Pied Flycatcher)
Western Palearctic. Four or five races. A single male.

AEGITHALIDAE

Aegithalos caudatus (Long-tailed Tit)
Palearctic. 19 races. A single bird.

PARIDAE

Parus palustris (Marsh Tit)
Western and eastern Palearctic and, marginally, Oriental. Up to 16 races described. Three birds.

Parus major (Great Tit)
Palearctic and Oriental. At least 33 races. A single female.

Parus caeruleus (Blue Tit)
Western Palearctic. 15 or 16 races. Two first-winters.

SITTIDAE

Sitta europaea (Nuthatch)
Palearctic and Oriental. Up to 28 races described. Three birds.

CERTHIIDAE

Certhia familiaris (Treecreeper)
Holarctic and, marginally, Oriental. Approximately 15 races described but variation is slight and clinal (apart from isolated races). A single bird.

EMBERIZIDAE

Miliaria calandra (Corn Bunting)
Western Palearctic. Two races. A single bird.

FRINGILLIDAE

Fringilla coelebs (Chaffinch)
Western and central Palearctic (introduced elsewhere). 14 races. Two males and a female.

Fringilla montifringilla (Brambling)
Northern Palearctic. Monotypic. Two males.

Carduelis spinus (Siskin)
Discontinuous, with populations in western and eastern Palearctic. Monotypic. A single female.

Carduelis carduelis (Goldfinch)
Western and central Palearctic and, marginally, Oriental (introduced elswhere). Up to 19 races described. A single adult.

Carduelis flammea (Redpoll)
Holarctic. Four races. Five birds. Three birds had originally been housed together.

Pinicola enucleator (Pine Grosbeak)
Holarctic. Ten or 11 races. The female-type had originally been cased with another, but the whereabouts of this bird is unknown.

Loxia curvirostra (Crossbill)
Holarctic and, marginally, Oriental. 20 races. A single female.

Coccothraustes coccothraustes (Hawfinch)
Palearctic. Five races. A single male and female.

PASSERIDAE

Passer domesticus (House Sparrow)
Cosmopolitan. 11 races. Two males.

Passer montanus (Tree Sparrow)
Palearctic and Oriental. Seven races. The two birds had previously been housed together.

STURNIDAE

Sturnus vulgaris (Starling)
Western and central Palearctic. 11 races. Two birds.

DICRURIDAE

Dicrurus paradiseus (Greater Racket-tailed Drongo)
Oriental. 14 races. A single adult.

CORVIDAE

Garrulus glandarius (Jay)
Palearctic and northern Oriental. Up to 36 races described. Two birds.

Pica pica (Magpie)
Holarctic, Oriental and Ethiopian (Saudi Arabia only). 13 races. Two adults and two first-years. The adults had originally been housed together and one of the first-years is depicted on a gamekeeper's gibbet.

Pyrrhocorax pyrrhocorax (Chough)
Southern Palearctic and northeastern Ethiopian. Seven or eight races. The two adults, which had previously been housed together, were taken at Inisboffin, Ireland in April 1888.

Corvus monedula (Jackdaw)
Western and central Palearctic. Four races. A single bird.

Corvus frugilegus (Rook)
Palearctic. Two races. A single adult.

Corvus corone (Carrion Crow)
Palearctic and, marginally, Oriental. Six races. Two birds. A bird of the nominate race is displayed looking into the empty nest of a ground nesting bird. The other, which had originally been cased, is a hybrid *C.c.corone* X *C.c.cornix* and was collected in Merioneth, North Wales on an unrecorded date.

Corvus corax (Raven)
Holarctic and northwest Neotropical. Seven or eight races. The single bird is still retaining most of its juvenile plumage but bears no collecting date.

EGGS

The Yorkshire Museum houses two major egg collections; one of which was built up by Museum staff in the mid-nineteenth century and the other which belonged to the late William Cooper.

Unlike the skins and mounts it has not been possible to apply a critical standard of verification and there remains much scope for the serious student of oology. For completeness, and as a service to ornithologists, however, the names appended to these eggs are listed below.

THE MUSEUM COLLECTION

The Yorkshire Museum Egg Collection is housed in 30 drawers (three cabinets) and consists of 1994 eggs belonging to 131 species.

PODICIPEDIFORMES

PODICIPEDIDAE

Tachybaptus ruficollis (Little Grebe)
Three clutches of 4 from Yorkshire.

Podiceps cristatus (Great Crested Grebe)
Clutches of 4 from Leicestershire and 2 from Surrey.

PROCELLARIIFORMES

PROCELLARIIDAE

Fulmarus glacialis (Fulmar)
A single egg from Highland, Scotland.

PELECANIFORMES

PHALACROCORACIDAE

Phalacrocorax carbo (Cormorant)
A clutch of 4 from Highland, Scotland and a clutch of 2 that is devoid of data.

Phalacrocorax aristotelis (Shag)
Two clutches of 3 (Highland, Scotland and Yorkshire).

CICONIIFORMES

ARADEIDAE

Botaurus stellaris (Bittern)
A clutch of 2 has no more precise locality data than "Russia".

Ardea cinerea (Grey Heron)
Three clutches (1 of 3 and 2 of 2). Clutch of 3 from Yorkshire and clutches of 2 from Highland, Scotland and Essex.

ANSERIFORMES

ANATIDAE

Anas crecca (Teal)
A clutch of 12 from Yorkshire belongs to the nominate race.

Anas platyrhynchos (Mallard)
Clutches of 12 and 2 from Yorkshire.

Somateria mollissima (Eider)
A clutch of 3 from Northumberland.

Mergus merganser (Goosander)
A clutch of 3 from northern Iceland.

FALCONIFORMES

PANDIONIDAE

Pandion haliaetus (Osprey)
A clutch of 2. The data label with these eggs has a question mark in front of the locality (Scotland).

ACCIPTRIDAE

Circus aeruginosus (Marsh Harrier)
A clutch of 4 from Pomerania (country not indicated).

Circus pygargus (Montagu's Harrier)
A clutch of 6 from Pomerania (country not indicated).

Accipiter nisus (Sparrowhawk)
Six clutches (1 of 6, 2 of 5, 2 of 4 and 1 of 3). Clutches of 6, 5 and 4 from Yorkshire, clutch of 5 from Lancashire whilst the others (4 and 3) are devoid of data.

Buteo buteo (Buzzard)
Clutches of 3 and 2 from Cumberland.

FALCONIDAE

Falco tinnunculus (Kestrel)
Five clutches (1 of 6, 2 of 5 and 2 of 3). All from Yorkshire with the exception of a clutch of 3 from Buckinghamshire.

Falco columbarius (Merlin)
Three clutches (1 of 6, 1 of 5 and 1 of 2) from Yorkshire.

Falco subbuteo (Hobby)
A clutch of 3 from near York, Yorkshire in 1888. Although the species was known to be a very occasional breeder in the county at this time, this record is not mentioned in Nelson's *The Birds of Yorkshire* (1907).

GALLIFORMES

TETRAONIDAE

Lagopus lagopus (Willow/Red Grouse)
A clutch of 5 from Yorkshire.

PHASIANIDAE

Alectoris rufa (Red-legged Partridge)
Two clutches (1 of 5 and 1 of 3) from Yorkshire.

Perdix perdix (Grey Partridge)
Two clutches (1 of 17 and 1 of 9) from Yorkshire.

GRUIFORMES

RALLIDAE

Crex crex (Corncrake)
A clutch of 5 from Yorkshire.

Gallinula chloropus (Moorhen)
Four clutches (2 of 8, 1 of 7 and 1 of 5) from Yorkshire.

Fulica atra (Coot)
Two clutches of 4 from Yorkshire.

OTIDAE

Otis tarda (Great Bustard)
A single egg from Scampston, Yorkshire in 1909 has a note attached indicating that it originated from an aviary.

CHARADRIIFORMES

HAEMATOPODIDAE

Haematopus ostralegus (Oystercatcher)
Five clutches (3 of 3 and 2 of 2). Two clutches of 3 and one of 2 from Cumberland, clutch of 3 from Northumberland and a clutch of 2 from Highland, Scotland.

BURHINIDAE

Burhinus oedicnemus (Stone-curlew)
Three clutches of 2 from Suffolk.

CHARADRIIDAE

Vanellus vanellus (Lapwing)
Seven clutches (5 of 4 , 1 of 3 and 1 of 2). All from Yorkshire with the exception of the clutch of 3 from Lincolnshire.

Pluvialis apricaria (Golden Plover)
Clutches of 4 from Lancashire and 3 from Yorkshire.

Charadrius hiaticula (Ringed Plover)
Four clutches (3 of 3 and 1 of 2). Clutches of 4 from Yorkshire and Cumberland (2) and a clutch of 2 from Lancashire.

SCOLOPACIDAE

Numenius phaeopus (Whimbrel)
A clutch of 2 from Pomerania (country not indicated). The locality data attached to these eggs, if they are correctly identified, is erroneous as the species does not breed in any of the Pomeranian countries.

Numenius arquata (Curlew)
A clutch of 4 from Yorkshire.

Tringa totanus (Redshank)
Ten clutches (8 of 4 and 2 of 3) from Yorkshire.

Actitis hypoleucos (Common Sandpiper)
Five clutches of 4 from Yorkshire.

Arenaria interpres (Turnstone)
A clutch of 4 that is devoid of data.

Scolopax rusticola (Woodcock)
Clutches of 4 from Tayside, Scotland and 3 from Yorkshire.

Gallinago gallinago (Snipe)
Seven clutches (1 of 5, 3 of 4, 2 of 3 and 1 of 2) from Yorkshire.

Caldris temminckii (Temminck's Stint)
A clutch of 4 from Swillington, Yorkshire. These are the eggs of the only English breeding attempt by this species. Fortunately the failure of these eggs to hatch was not caused by human interference and a label, written by Dr E.W. Taylor on 15th September 1951, elucidates:

"The only known clutch of the eggs of Temminck's Stint, Calidris temminckii, from a locality near Leeds. Every effort was made to protect the bird and nest but the former was killed by a Weasel or Rat, July 8, 1951. The eggs were presented by Mr W. Bennett on behalf of the Leeds Bird Watchers Club. The eggs were fertile and within a few days of hatching"

Normally breeding no nearer than Norway this record is quite extraordinary.

Calidris alpina (Dunlin)
Clutches of 4 from Iceland and 3 from Lancashire.

Philomacus pugnax (Ruff)
A clutch of 4 from Yorkshire.

LARIDAE

Larus canus (Common Gull)
A clutch of 3 from Highland, Scotland.

Larus argentatus (Herring Gull)
Two clutches of 3 from Scotland.

Larus fuscus (Lesser Black-backed Gull)
Clutches of 3 and 2 from Scotland.

Larus ridibundus (Black-headed Gull)
13 clutches (1 of 4, 10 of 3 and 2 of 2). All from Cumberland with the exception of a clutch of 3 from Yorkshire.

Rissa tridactyla (Kittiwake)
22 clutches (6 of 3, 3 of 2 and 13 of 1) from Bempton, Yorkshire.

STERNIDAE

Sterna hirundo (Common Tern)
A clutch of 3 from Anglesey, Wales.

Sterna paradisaea (Arctic Tern)
Clutches of 3 and 2 from Highland, Scotland.

Sterna hirundo / Sterna paradisaea (Common / Arctic Tern)
Ten clutches (3 of 3 and 7 of 2) are lacking data and have a label attached indicating that specific determination is uncertain.

Sterna dougallii (Roseate Tern)
A clutch of 2 from the Welsh coast.

Sterna albifrons (Little Tern)
Clutches of 3 and 2 from Yorkshire.

Sterna sandvicensis (Sandwich Tern)
Ten clutches (8 of 2 and 2 of 1). All from Cumberland with the exception of two clutches of 2 and one of 1 from Northumberland.

ALCIDAE

Alca torda (Razorbill)
The 42 eggs originate from Bempton, Yorkshire and Ailsa Craig, Scotland but no distinction is indicated.

Uria aalge (Guillemot)
The 65 eggs originate from Ailsa Craig, Scotland and Bempton, Yorkshire but no distinction is indicated.

Fratercula arctica (Puffin)
12 clutches of single eggs. All from the Farne Islands, Northumberland with the exception of 1 from Highland, Scotland.

COLUMBIFORMES

COLUMBIDAE

Columba livia (Rock Dove)
A clutch of 2 from Bempton, Yorkshire.

Columba oenas (Stock Dove)
A clutch of 2 from Yorkshire.

Columba palumbus (Woodpigeon)
A clutch of 2 from Yorkshire.

Streptopelia turtur (Turtle Dove)
Three clutches of 2 (two from Yorkshire and one from Kent).

CUCULIFORMES

CUCULIDAE

Cuculus canorus (Cuckoo)
A single egg of this species can be found with the following host species' eggs:

Anthus pratensis Meadow Pipit. Two clutches (1 of 4 and 1 of 2) are devoid of locality data.
Prunella modularis Dunnock. A clutch of 4 is devoid of locality data.

Additionally there are three eggs of this species which are said to originate from the nests of Dunnock *Prunella modularis* containing 2 eggs (Yorkshire and no locality) and 1 egg (Kent), but the host species' eggs have not been retained.

STRIGIFORMES

TYTONIDAE

Tyto alba (Barn Owl)
A clutch of 4 from Yorkshire.

STRIGIDAE

Athene noctua (Little Owl)
Two clutches of 4 (Yorkshire and Hereford & Worcester).

Strix aluco (Tawny Owl)
Three clutches (2 of 4 and 1 of 3) from Yorkshire.

Asio otus (Long-eared Owl)
Four clutches (1 of 5 and 3 of 3). Clutches of 5 and 3 from Yorkshire and clutches of 3 from Northumberland and Lincolnshire.

CAPRIMULGIFORMES

CAPRIMULGIDAE

Caprimulgus europaeus (Nightjar)
Three clutches of 2 (two from Yorkshire and one from Berkshire).

APODIFORMES

APODIDAE

Apus apus (Swift)
Three clutches of 5 from Yorkshire.

CORACIIFORMES

ALCEDINIDAE

Alcedo atthis (Kingfisher)
A clutch of 7 from Yorkshire.

UPUPIDAE

Upupa epops (Hoopoe)
A clutch of 3 from southern Spain.

PICIFORMES

PICIDAE

Jynx torquilla (Wryneck)
Two clutches (1 of 8 and 1 of 6) from Kent.

PASSERIFORMES

ALAUDIDAE

Lullula arborea (Woodlark)
Two clutches of 4 (Hampshire and France).

Alauda arvensis (Skylark)
Three clutches of 4 (two from Yorkshire the other devoid of data).

HIRUNDINIDAE

Riparia riparia (Sand Martin)
Four clutches (1 of 5 and 3 of 4) from Yorkshire.

Hirundo rustica (Swallow)
Six clutches (3 of 5, 2 of 4 and 1 of 3) from Yorkshire.

Delichon urbica (House Martin)
A clutch of 4 from Nottingham.

MOTACILLIDAE

Motacilla flava (Yellow Wagtail)
Five clutches (1 of 6, 3 of 5 and 1 of 4) from Yorkshire. A clutch of 5 from Wiggington on 13th June 1906 is said to belong to the nominate race.

Motacilla cinerea (Grey Wagtail)
A clutch of 5 from Northumberland.

Motacilla alba (Pied Wagtail)
Four clutches (3 of 6 and 1 of 4) from Yorkshire.

Anthus pratensis (Meadow Pipit)
Eight clutches (3 of 5 and 5 of 4). Two clutches of 5 and two of 4 from France, a clutch of 5 and two of 4 from Yorkshire and a clutch of 4 from Northumberland. (See *Cuculus canorus* (Cuckoo)).

Anthus trivialis (Tree Pipit)
13 clutches (2 of 6, 3 of 5, 7 of 4 and 1 of 3). All from Yorkshire with the exception of a clutch of 4 from Northumberland.

Anthus petrosus (Rock Pipit)
A clutch of 5 from Northumberland and a clutch of 4 from Ailsa Craig, Scotland.

LANIIDAE

Lanius collurio (Red-backed Shrike)
Four clutches (1 of 6, 2 of 5 and 1 of 1). All from France with the exception of the single from Buckinghamshire.

CINCLIDAE

Cinclus cinclus (Dipper)
A clutch of 3 from Yorkshire.

TROGLODYTIDAE

Troglodytes troglodytes (Wren)
Four clutches (2 of 7 and 2 of 6). All from Yorkshire with the exception of a clutch of 6 from Kent.

PRUNELLIDAE

Prunella modularis (Dunnock)
Two clutches (1 of 5 and 1 of 4) from Yorkshire. (See *Cuculus canorus* (Cuckoo)).

TURDIDAE

Erithacus rubecula (Robin)
Nine clutches (1 of 7, 4 of 6, 3 of 5 and 1 of 4). Clutches of 6, 6, 5, 5 and 4 from Yorkshire, clutches of 7, 6 and 5 from Kent and a clutch of 6 that is devoid of data.

Luscinia megarhynchos (Nightingale)
Five clutches (4 of 5 and 1 of 4). All from France with the exception of the clutch of 4 from Sussex.

Phoenicurus phoenicurus (Redstart)
A clutch of 6 from Yorkshire.

Saxicola rubetra (Whinchat)
Four clutches (2 of 6 and 2 of 5). All from Yorkshire with the exception of a clutch of 6 from northern France.

Saxicola torquata (Stonechat)
Two clutches (1 of 6 from France and 1 of 4 from Sussex).

Oenanthe oenanthe (Wheatear)
Three clutches (1 of 7, 1 of 5 and 1 of 4). Clutch of 7 from Yorkshire; the others from northern France.

Turdus torquatus (Ring Ouzel)
Two clutches of 4 from Yorkshire.

Turdus merula (Blackbird)
11 clutches (1 of 6, 2 of 5, 6 of 4, 1 of 3 and 1 of 1). All from Yorkshire with the exception of a clutch of 4 from Kent. The single egg, which is said to belong to this species, was taken at Haxby in June 1952 and only measures 10mm X 9mm (the average size is 29.4mm X 21.7mm).

Turdus philomelos (Song Thrush)
Six clutches (1 of 5, 4 of 4 and 1 of 3). Clutches of 5, 4 and 4 from Yorkshire, clutches of 4 and 3 from Kent and a clutch of 4 that is devoid of data.

142

Turdus viscivorus (Mistle Thrush)
Six clutches (1 of 5, 4 of 4 and 1 of 2) from Yorkshire.

SYLVIIDAE

Locustella naevia (Grasshopper Warbler)
A clutch of 4 from Yorkshire.

Acrocephalus schoenobaenus (Sedge Warbler)
Eight clutches (1 of 6, 3 of 5, 3 of 4 and 1 of 3) from Yorkshire.

Acrocephalus scirpaceus (Reed Warbler)
Three clutches of 4 (two from Sussex and one from France).

Sylvia borin (Garden Warbler)
Six clutches (1 of 5, 2 of 4, 2 of 3 and 1 of 2). Clutches of 5, 4, 3 and 2 from Kent, clutch of 4 from Yorkshire and a clutch of 3 from Northumberland.

Sylvia atricapilla (Blackcap)
Three clutches (1 of 4 and 2 of 2). Clutches of 4 and 2 from Yorkshire and clutch of 2 from Kent.

Sylvia communis (Whitethroat)
15 clutches (2 of 6, 10 of 5 and 3 of 4). All from Yorkshire with the exception of a clutch of 5 from Kent.

Sylvia curruca (Lesser Whitethroat)
Three clutches (1 of 5 and 2 of 4). Clutches of 5 and 4 from Yorkshire and clutch of 4 from Kent.

Sylvia undata (Dartford Warbler)
A clutch of 4 from Hampshire.

Phylloscopus trochilus (Willow Warbler)
Six clutches (3 of 6 and 3 of 5). All from Yorkshire with the exception of a clutch of 6 from Kent.

Phylloscopus colybita (Chiffchaff)
A clutch of 6 from Yorkshire.

143

Phylloscopus sibilatrix (Wood Warbler)
Three clutches (1 of 6 and 2 of 4). Clutch of 6 from Kent and clutches of 4 from Yorkshire and Northumberland.

MUSCICAPIDAE

Muscicapa striata (Spotted Flycatcher)
Six clutches (3 of 5, 2 of 4 and 1 of 3). All from Yorkshire with the exception of a clutch of 4 from Sussex.

AEGITHALIDAE

Aegithalos caudatus (Long-tailed Tit)
A clutch of 8 from Sussex.

PARIDAE

Parus palustris (Marsh Tit)
Two clutches (1 of 8 from Sussex and 1 of 4 from Yorkshire).

Parus montanus (Willow Tit)
A clutch of 8 from Leicestershire.

Parus ater (Coal Tit)
A clutch of 7 from Kent.

Parus major (Great Tit)
Three clutches (1 of 10, 1 of 9 and 1 of 4). Clutch of 10 from Lincolnshire; the others from Yorkshire.

Parus caeruleus (Blue Tit)
Clutches of 9 and 8 from Yorkshire.

CERTHIIDAE

Certhia familiaris (Treecreeper)
Three clutches (2 of 6 from Yorkshire and 1 of 3 from Kent).

EMBERIZIDAE

Miliaria calandra (Corn Bunting)
Three clutches (1 of 5, 1 of 4 and 1 of 3). Clutches of 5 and 4 from France the other from Yorkshire.

Emberiza citrinella (Yellowhammer)
Eight clutches (6 of 4 and 2 of 3). All from Yorkshire with the exception of a clutch of 4 that is devoid of data. A clutch of 3 from Sutton on Forest on 22nd June 1932 is completely white and lacks any pigmentation.

Emberiza schoeniclus (Reed Bunting)
Four clutches (3 of 5 and 1 of 4). All from Yorkshire with the exception of a clutch of 5 from France.

FRINGILLIDAE

Fringilla coelebs (Chaffinch)
Seven clutches (3 of 5, 2 of 4, 1 of 3 and 1 of 2). All from Yorkshire with the exception of a clutch of 3 from Durham. One of the eggs in the Durham clutch is but a third of the size of the other two.

Carduelis chloris (Greenfinch)
Four clutches (1 of 6, 2 of 5 and 1 of 4) from Yorkshire. A clutch of 5 from West Cottingwith on 7th May 1922 is completely white and lacks any pigmentation.

Carduelis flammea (Redpoll)
Three clutches (1 of 5, 1 of 3 and 1 of 2) from Yorkshire.

Carduelis cannabina (Linnet)
Three clutches (2 of 5 and 1 of 3) from Yorkshire.

Pyrrhula pyrrhula (Bullfinch)
Three clutches of 4 from Yorkshire.

Coccothraustes coccothraustes (Hawfinch)
Two clutches (one of 4 from Bishopstone and one of 3 from Kent). The county of origin of the clutch of 4 remains a mystery as there are several localities in the British Isles bearing this name.

PASSERIDAE

Passer domesticus (House Sparrow)
Six clutches (2 of 6, 2 of 5 and 2 of 4). All from Yorkshire with the exception of a clutch of 6 that is devoid of data.

Passer montanus (Tree Sparrow)
Five clutches (4 of 5 and 1 of 4). All from Yorkshire with the exception of a clutch of 5 from Norfolk.

STURNIDAE

Sturnus vulgaris (Starling)
Three clutches (1 of 7, 1 of 5 and 1 of 4) from Yorkshire.

CORVIDAE

Garrulus glandarius (Jay)
Two clutches (1 of 6 and 1 of 5) from Yorkshire.

Pica pica (Magpie)
Six clutches (1 of 7, 2 of 6 and 3 of 5) from Yorkshire.

Pyrrhocorax pyrrhocorax (Chough)
A clutch of three from Blaskett Island, County Kerry, Ireland.

Corvus monedula (Jackdaw)
Five clutches (3 of 5 and 2 of 4) from Yorkshire.

Corvus frugilegus (Rook)
Seven clutches (5 of 5 and 2 of 4) from Yorkshire.

Corvus corone (Carrion Crow)
11 clutches (1 of 7, 2 of 6, 1 of 5, 5 of 4, 1of 3 and 1 of 2). All from Yorkshire with the exception of a clutch of 6 from France.

THE WILLIAM COOPER COLLECTION

The William Cooper Egg Collection is housed in a 72 drawer cabinet and consists of 4,350 eggs of 309 species that were mainly collected towards the end of the nineteenth century.

The vagueness of locality data in the pages which follow i.e. southern Russia and Asia Minor, is due to the limited data attached to the clutches of eggs within the collection.

GAVIIFORMES

GAVIIDAE

Gavia stellata (Red-throated Diver)
Four clutches (2 of 2 and 2 of 1) from northern Scotland.

Gavia arctica (Black-throated Diver)
Four clutches (3 of 2 and 1 of 1) from northern Scotland.

Gavia immer (Great Northern Diver)
Two clutches of 2 from Iceland.

PODICIPEDIFORMES

PODICIPEDIDAE

Tachybaptus ruficollis (Little Grebe)
Two clutches (1 of 4 and 1 of 3) from Yorkshire.

Podiceps grisegena (Red-necked Grebe)
A clutch of 4 from Finland.

Podiceps cristatus (Great Crested Grebe)
A clutch of 5 from Ireland.

Podiceps auritus (Slavonian Grebe)
A clutch of 6 from northern Iceland.

Podiceps nigricollis (Black-necked Grebe)
A clutch of 5 from Ireland (June 1893). If the data attached to these eggs is correct, and there is no evidence to suggest otherwise, then the record predates the first recorded breeding of this species in Ireland by more than 20 years!

PROCELLARIIFORMES

PROCELLARIIDAE

Fulmarus glacialis (Fulmar)
Three single eggs (two from St. Kilda, Shetland, Scotland; the other from Speeton, Yorkshire).

Bulweria bulwerii (Bulwer's Petrel)
A single egg from Madeira.

Puffinus gravis (Great Shearwater)
A single egg from Chatham Island, New Zealand. The locality data attached to this egg, assuming it is correctly identified, is erroneous as the species does not breed in New Zealand.

Puffinus griseus (Sooty Shearwater)
A single egg from Snares Island, New Zealand.

Puffinus puffinus (Manx Shearwater)
Three single eggs from Orkney, Scotland.

Puffinus assimilis (Little Shearwater)
A single egg from Porto Santo, Madeira.

HYDROBATIDAE

Pelagodroma marina (White-faced Storm Petrel)
A single egg from Tom Thumb Island, New South Wales, Australia.

Hydrobates pelagicus (Storm Petrel)
Two single eggs from Orkney, Scotland.

Oceanodroma leucorhoa (Leach's Petrel)
Two single eggs from Duck Island, Massachusetts, North America.

PELECANIFORMES

SULIDAE

Morus bassanus (Gannet)
A clutch of 2 from Orkney, Scotland.

PHALACROCORACIDAE

Phalacrocorax carbo (Cormorant)
A clutch of 3 from Wales.

Phalacrocorax aristotelis (Shag)
Two clutches of 3 from northern Scotland.

CICONIIFORMES

ARADEIDAE

Botaurus stellaris (Bittern)
A clutch of 5 from southern Russia.

Botaurus lentiginosus (American Bittern)
A clutch of 5 from Tennessee, North America.

Nycticorax nycticorax (Night Heron)
A clutch of 5 from Sweden. The locality data attached to these eggs, assuming they are correctly identified, is erroneous as the species does not breed in Sweden.

Ardeola ralloides (Squacco Heron)
Two clutches (1 of 4 and 1 of 3) from southern Russia (the clutch of 4 from the mouth of the Volga River).

Bubulcus ibis (Cattle Egret)
A clutch of 5 from southern Russia.

Butorides virescens (Green Heron)
A clutch of 4 from Tennessee, North America.

Egretta garzetta (Little Egret)
A clutch of 3 from southern Russia.

Egretta alba (Great White Egret)
A clutch of 4 from southern Florida.

Ardea purpurea (Purple Heron)
A clutch of 3 from southern Russia.

Ardea cinerea (Grey Heron)
A clutch of 4 from Ireland.

CICONIIDAE

Ciconia nigra (Black Stork)
A clutch of 3 has indecipherable locality data.

Ciconia ciconia (White Stork)
A clutch of 3 from Holland.

THRESKIORNITHIDAE

Plegadis falcinellus (Glossy Ibis)
A clutch of 4 from Yugoslavia.

Platalea leucorodia (Spoonbill)
A clutch of 3 is devoid of data.

PHOENICOPTERIDAE

Phoenicopterus ruber (Greater Flamingo)
A single egg from southern Spain.

ANSERIFORMES

ANATIDAE

Cygnus olor (Mute Swan)
A clutch of 2 from northern Scotland.

Cygnus cygnus (Whooper Swan)
A single egg from Iceland.

Anser fabalis (Bean Goose)
A clutch of 2 is devoid of data.

Anser anser (Greylag Goose)
A single egg from northern Scotland.

Anser brachyrhynchus (Pink-footed Goose)
A clutch of 2 from Iceland.

Anser albifrons (White-fronted Goose)
Two clutches (2 and 1) are devoid of data.

Tadorna ferruginea (Ruddy Shelduck)
A clutch of 3 from southern Russia.

Tadorna tadorna (Shelduck)
A clutch of 2 from Highland, Scotland.

Anas penelope (Wigeon)
A clutch of 9 from northern Scotland.

Anas americana (American Wigeon)
A clutch of 4 from north-west Canada.

Anas strepera (Gadwall)
A clutch of 6 from Iceland.

Anas crecca (Teal)
Two clutches (1 of 9 from northern Scotland and 1 of 4 from north-west America). The clutch from Scotland will belong to the nominate race whilst the American clutch will be that of the race *carolinensis.*

Anas platyrhynchos (Mallard)
A clutch of 6 from northern Scotland. A clutch of 9 recently added eggs (1972) are said to be those of *"Cayuga Mallard"* and were acquired at Heslington, Yorkshire.

Anas acuta (Pintail)
A clutch of 5 is devoid of data.

Anas querquedula (Garganey)
A clutch of 7 from northern Iceland. The locality data attached to these eggs, assuming they are correctly identified, is erroneous as the species does not breed in Iceland.

Anas discors (Blue-winged Teal)
A clutch of 11 from Manitoba, Canada.

Anas clypeata (Shoveler)
A clutch of 6 has indecipherable locality data.

Netta rufina (Red-crested Pochard)
A clutch of 6 from southern Russia.

Aythya ferina (Pochard)
A clutch of 2 from Scotland.

Aythya nyroca (Ferruginous Duck)
A clutch of 4 from Yugoslavia.

Aythya fuligula (Tufted Duck)
A clutch of 2 is devoid of locality data.

Aythya marila (Scaup)
A clutch of 6 from northern Iceland.

Somateria mollissima (Eider)
A clutch of 5 from northern Scotland.

Somateria spectabilis (King Eider)
A clutch of 2 from Alaska.

Histrionicus histrionicus (Harlequin)
A clutch of 5 from northern Iceland.

Clangula hyemalis (Long-tailed Duck)
A clutch of 6 from Iceland.

Melanitta nigra (Common Scoter)
A clutch of 12 from northern Scotland.

Melanitta fusca (Velvet Scoter)
Two clutches (1 of 3 from Sweden and 1 of 1 from Lapland).

Bucephala clangula (Goldeneye)
A clutch of 6 from Iceland. The locality data attached to these eggs,
- assuming they are correctly identified, is erroneous as the species does
not breed in Iceland. Its congener, *B. islandica* (Barrow's Goldeneye)
does, however!

Mergus cucullatus (Hooded Merganser)
A clutch of 5 from North America.

Mergus serrator (Red-breasted Merganser)
A clutch of 7 from northern Scotland.

Mergus merganser (Goosander)
A clutch of 6 from Finland.

FALCONIFORMES

PANDIONIDAE

Pandion haliaetus (Osprey)
Three clutches (1 of 3, and 2 of 1) from America.

ACCIPITRIDAE

Pernis apivorus (Honey Buzzard)
Two clutches of 2 (Germany and Finland).

Milvus migrans (Black Kite)
Three clutches (2 of 4 and 1 of 3). Clutches of 4 from Russia and 3 from Switzerland. The other clutch of 4 is devoid of data.

Milvus milvus (Red Kite)
Three clutches of 3 (Norway, Germany and Spain).

Haliaeetus albicilla (White-tailed Sea Eagle)
Two clutches of 3 from southern Russia.

Neophron percnopterus (Egyptian Vulture)
Three clutches of 2 (2 from Asia Minor and 1 from southern Spain).

Gyps fulvus (Griffon Vulture)
Two clutches of 2 (north Africa and southern Spain).

Circus aeruginosus (Marsh Harrier)
Two clutches (4 and 3) from southern Russia.

Circus cyaneus (Hen Harrier)
Three clutches (1 of 4 and 2 of 2) from Orkney, Scotland.

Circus pygargus (Montagu's Harrier)
A clutch of 5 from Germany.

Accipiter gentilis (Goshawk)
A clutch of 3 from Germany.

153

Accipiter nisus (Sparrowhawk)
16 clutches (1 of 7, 4 of 6, 8 of 5, 1 of 4 and 2 of 3) from Yorkshire.

Buteo buteo (Buzzard)
Three clutches (2 of 3 from Germany and 1 of 2 from Wales).

Buteo lagopus (Rough-legged Buzzard)
Three clutches (1 of 2, 1 of 3 and 1 of 4). Clutches of 2 and 3 from Lapland and clutch of 4 from Sweden.

Aquila clanga (Great Spotted Eagle)
Three clutches of 2 from southern Russia.

Aquila chrysaetos (Golden Eagle)
Three clutches (2 of 2 and 1 of 1) from Scotland.

FALCONIDAE

Falco naumanni (Lesser Kestrel)
A clutch of 5 from Turkey.

Falco tinnunculus (Kestrel)
29 clutches (6 of 6, 12 of 5, 8 of 4 and 3 of 3). All from Yorkshire with the exception of a clutch of 5 and 2 of 4 from Scotland.

Falco vespertinus (Red-footed Falcon)
A clutch of 4 from Hungary.

Falco columbarius (Merlin)
11 clutches (6 of 5, 4 of 4 and 1 of 1). All from Scotland with the exception of a clutch of 4 from Yorkshire.

Falco subbuteo (Hobby)
Two clutches (1 of 3 and 1 of 6). Clutch of 3 from Asia Minor, the others from Germany. The clutch of 6 was in the cabinet standing under the name of Red-footed Falcon *Falco vespertinus.*

Falco rusticolus (Gyrfalcon)
Five clutches (4 of 4 and 1 of 2). Clutches of 4 from Greenland (2), Iceland (1) and Sweden(1). Clutch of 2 from Iceland. Due to persecution the species no longer breeds in Sweden!

Falco peregrinus (Peregrine)
Four clutches (2 of 4, 1 of 2 and 1 of 1). Clutch of 2 from Sweden, the others from Scotland.

GALLIFORMES

TETRAONIDAE

Lagopus lagopus (Willow/Red Grouse)
Two clutches (1 of 8 and 1 of 5) from northern Scotland.

Lagopus mutus (Ptarmigan)
A clutch of 7 from Iceland.

Tetrao tetrix (Black Grouse)
A clutch of 8 from Wales.

Tetrao urogallus (Capercaillie)
A clutch of 4 from Sweden.

PHASIANIDAE

Alectoris rufa (Red-legged Partridge)
A clutch of 12 from Yorkshire.

Coturnix coturnix (Quail)
Two clutches of 8 (Germany and Canary Isles).

GRUIFORMES

GRUIDAE

Grus grus (Crane)
A single egg is devoid of data.

Anthropoides virgo (Demoiselle Crane)
A clutch of 2 from the Volga River, Russia.

RALLIDAE

Rallus aquaticus (Water Rail)
A clutch of 10 from Holland.

Crex crex (Corncrake)
Six clutches (1 of 10, 1 of 9, 1 of 8 and 3 of 7). All from Yorkshire with the exception of the clutch of 10 from Highland, Scotland.

Porzana parva (Little Crake)
A clutch of 10 from southern Russia.

Porzana pusilla (Baillon's Crake)
A clutch of 8 from Hungary.

Porzana porzana (Spotted Crake)
A clutch of 7 from Hungary.

Gallinula chloropus (Moorhen)
A clutch of 8 from Yorkshire.

Porphyrio porphyrio (Purple Gallinule)
A clutch of 5 from southern Russia.

Fulica atra (Coot)
A clutch of 2 from Orkney, Scotland.

OTIDAE

Tetrax tetrax (Little Bustard)
A clutch of 2 from southern Spain.

Otis tarda (Great Bustard)
A clutch of 2 from Mala, N.W. Rhodesia, Africa. As this species does not breed in Africa, the locality data attached to these eggs is erroneous or they do not belong to this species. If one assumes that these eggs belong to *Otis tarda*, and the measurements (72mm X 53mm and 71mm X 53mm) fall within the range for this species, then the locality data is erroneous. On the other hand, as the eggs fall within the size range for Denham's Bustard *Neotis denhami* (the only species found in Rhodesia which lays eggs of a comparable size), it is possible they belong to this species. In the days when it was legal to collect eggs, many dealers, some unscrupulous, would palm off the unsuspecting buyer with wrongly identified and / or fictitious data. It would appear that this may have happened in this instance and the solution to the eggs' true identity may remain a mystery forever!

Chlamydotis undulata (Houbara Bustard)
A clutch of 3 from Russia.

CHARADRIIFORMES

HAEMATOPODIDAE

Haematopus ostralegus (Oystercatcher)
Three clutches of 3 from Highland, Scotland.

RECURVIROSTRIDAE

Himantopus himantopus (Black-winged Stilt)
Two clutches of 4 from Turkey.

Recurvirostra avosetta (Avocet)
Two clutches of 4 from southern Spain.

BURHINIDAE

Burhinus oedicnemus (Stone-curlew)
Five clutches of 2 from Yorkshire (1), Norfolk (1) and the Canary Isles
(3). The Yorkshire clutch was taken in 1908. Formerly a common breeder
on the Yorkshire Wolds (until the middle of the nineteenth century), the
species was only nesting in very small numbers by the time of this clutch
and no longer breeds in the county.

GLAREOLIDAE

Cursorius cursor (Cream-coloured Courser)
Two clutches of 2 from the Canary Isles.

Glareola pratincola (Collared Pratincole)
A clutch of 2 from Asia Minor.

CHARADRIIDAE

Vanellus vanellus (Lapwing)
Five clutches (1 of 4 and 4 of 3). Three clutches of 3 from northern
Scotland, the others from Yorkshire.

Vanellus gregarius (Sociable Plover)
A clutch of 4 from the Crimea, Russia.

Pluvialis apricaria (Golden Plover)
Four clutches (3 of 4 and 1 of 2). Two clutches of 4 from Yorkshire, the others from northern Scotland.

Charadrius hiaticula (Ringed Plover)
Three clutches (2 of 4 and 1 of 3). Clutches of 4 and 3 from Highland, Scotland the others from Belfast, Ireland.

Charadrius dubius (Little Ringed Plover)
A clutch of 3 from an untraced locality.

Charadrius vociferus (Killdeer)
A clutch of 4 from Canada.

Charadrius alexandrinus (Kentish Plover)
Two clutches of 3. One from Germany, the other devoid of data.

Charadrius morinellus (Dotterel)
Two clutches of 3 from Egypt, Africa. The locality data attached to these eggs, assuming they are correctly identified, is erroneous as the species does not breed in Africa.

SCOLOPACIDAE

Limosa limosa (Black-tailed Godwit)
A clutch of 2 from Iceland.

Limosa lapponica (Bar-tailed Godwit)
A clutch of 2 from Arkhangel'sk, Russia.

Numenius phaeopus (Whimbrel)
A clutch of 4 from Finland.

Numenius arquata (Curlew)
Seven clutches of 4 from northern Scotland (4) and Yorkshire (3).

Bartramia longicauda (Upland Sandpiper)
A clutch of 4 from South Dakota, North America.

Tringa totanus (Redshank)
Five clutches of 4 from Highland, Scotland.

Tringa nebularia (Greenshank)
Two clutches of 3 from northern Scotland.

Tringa glareola (Wood Sandpiper)
Three clutches (1 of 4 and 2 of 3) from Lapland.

Actitis hypoleucos (Common Sandpiper)
Four clutches of 4. All from northern Scotland with the exception of a clutch from Yorkshire.

Arenaria interpres (Turnstone)
A clutch of 4 from Sweden.

Phalaropus lobatus (Red-necked Phalarope)
Two clutches of 4 from Iceland.

Phalaropus fulicarius (Grey Phalarope)
A clutch of 4 from Iceland.

Scolopax rusticola (Woodcock)
Five clutches (3 of 4, 1 of 3 and 1 of 1) from Yorkshire.

Gallinago gallinago (Snipe)
Five clutches (4 of 4 and 1 of 3). Clutches of 4 from Yorkshire (3) and Suffolk (1). Clutch of 3 from Highland, Scotland. One of the eggs in a Yorkshire clutch is pure white, the others being normal.

Lymnocryptes minima (Jack Snipe)
Two clutches (1 of 4 from Sweden and 1 of 3 from Finland).

Calidris temminckii (Temminck's Stint)
A clutch of 4 from the Gulf of Bothnia (Sweden or Finland).

Calidris maritima (Purple Sandpiper)
A clutch of 2 from Iceland.

Calidris alpina (Dunlin)
Two clutches of 4 from northern Scotland.

Limicola falcinellus (Broad-billed Sandpiper)
Two clutches (1 of 4 and 1 of 3) from northern Sweden.

Philomacus pugnax (Ruff)
Two clutches of 3 from Lapland.

STERCORARIIDAE

Stercorarius skua (Great Skua)
A clutch of 2 from Iceland.

Stercorarius parasiticus (Arctic Skua)
Three clutches of 2 from Shetland, Scotland.

Stercorarius longicaudus (Long-tailed Skua)
A clutch of 2 from Norway.

LARIDAE

Larus canus (Common Gull)
Five clutches (1 of 4 and 4 of 3) from northern Scotland.

Larus argentatus (Herring Gull)
Four clutches (3 of 2 and 1 of 1) from northern Scotland.

Larus fuscus (Lesser Black-backed Gull)
Four clutches (3 of 3 and 1 of 2) from northern Scotland.

Larus marinus (Great Black-backed Gull)
Four clutches (1 of 3, 2 of 2 and 1 of 1) from northern Scotland.

Larus hyperboreus (Glaucous Gull)
A clutch of 3 from the Canary Isles. The locality data attached to these eggs, assuming they are correctly identified, is erroneous. The nearest breeding station to the Canary Isles for this species is southern Greenland!

Larus glaucoides (Iceland Gull)
A clutch of 2 from Greenland.

Larus ichthyaetus (Great Black-headed Gull)
A clutch of 3 from the Volga River, Russia.

Larus ridibundus (Black-headed Gull)
Five clutches (3 of 4, 1 of 3 and 1 of 2). All from northern Scotland with the exception of the clutch of 2 from Scarborough, Yorkshire.

Larus philadelphia (Bonaparte's Gull)
A single egg is devoid of data.

Larus minutus (Little Gull)
A clutch of 3 from Estonia.

Rissa tridactyla (Kittiwake)
Two clutches of 2 from Bempton, Yorkshire.

STERNIDAE

Chlidonias hybrida (Whiskered Tern)
A clutch of 3 from southern Russia.

Chlidonias leucoptera (White-winged Black Tern)
A clutch of 3 from Hungary.

Chlidonias nigra (Black Tern)
Four clutches of 3 from North America.

Gelochelidon nilotica (Gull-billed Tern)
A clutch of 3 from Texas, North America.

Sterna caspia (Caspian Tern)
A clutch of 3 from Texas, North America.

Sterna hirundo (Common Tern)
Two clutches (1 of 2 from northern Scotland and 1 of 1 from northern Iceland). The locality data attached to the clutch of 1, assuming it is correctly identified, is erroneous as the species does not breed in Iceland. Its congener, *Sterna paradisaea* (Arctic Tern), does, however!

Sterna paradisaea (Arctic Tern)
17 clutches (1 of 4, 10 of 3 and 6 of 2) from northern Scotland.

Sterna dougallii (Roseate Tern)
Three clutches (1 of 3 and 1 of 2 from North America and 1 of 2 from the Bahamas, West Indies).

Sterna fuscata (Sooty Tern)
Six clutches of single eggs (four from the Bahamas, West Indies; one from Texas, North America and one devoid of locality data). The locality data attached to the egg from Texas, assuming it is correctly identified, is erroneous as the species does not breed in this American State.

Sterna albifrons (Little Tern)
Two clutches of 3 from England.

Sterna sandvicensis (Sandwich Tern)
A clutch of 3 from the Scilly Isles.

Anous stolidus (Common Noddy)
Four clutches of single eggs from the Bahamas, West Indies.

ALCIDAE

Alle alle (Little Auk)
A single egg is devoid of data.

Alca torda (Razorbill)
Three clutches of single eggs from Speeton, Yorkshire.

Uria lomvia (Brunnich's Guillemot)
Five clutches of single eggs (3 from Labrador, Canada and 2 from Iceland).

Uria aalge (Guillemot)
The 54 eggs, which bear no collecting data, show the very variable coloration produced by this species.

Cepphus grylle (Black Guillemot)
Five clutches (4 of 2 and 1 of 1) from northern Scotland.

Fratercula arctica (Puffin)
Five clutches (1 of 2 and 4 of 1) from the east coast of Yorkshire. This species normally only lays a single egg, clutches of 2 are a rare occurrence.

COLUMBIFORMES

PTEROCLIDIDAE

Syrrhaptes paradoxus (Pallas's Sandgrouse)
A clutch of 3 from Ubekskaya, Russia.

COLUMBIDAE

Columba livia (Rock Dove)
Three clutches (2 of 2 and 1 of 1) from Scotland.

Columba oenas (Stock Dove)
Two clutches of 2 from Yorkshire.

Columba palumbus (Woodpigeon)
A clutch of 2 from Yorkshire.

CUCULIFORMES

CUCULIDAE

Clamator glandarius (Great Spotted Cuckoo)
A single egg along with six of those belonging to its main host species,
Magpie *Pica pica*.

Cuculus canorus (Cuckoo)
A single egg of this species can be found with the following host species'
eggs:

Alauda arvensis Skylark. Four eggs from Yorkshire.
Motacilla cinerea Grey Wagtail. Five eggs from Yorkshire.
Anthus pratensis Meadow Pipit. Three eggs from Yorkshire.
Anthus trivialis Tree Pipit. Two clutches of 4 eggs from Yorkshire.
Prunella modularis Dunnock. Four clutches (2 of 4 and 2 of 2) from
Yorkshire.
Erithacus rubecula Robin. Five eggs from Yorkshire.
Luscinia megarhynchos Nightingale. Three eggs from Hertfordshire.
Locustella naevia Grasshopper Warbler. Four eggs from Northumberland.
Acrocephalus schoenobaenus Sedge Warbler. Five eggs from Yorkshire.
Acrocephalus scirpaceus Reed Warbler. Three eggs from Berkshire.
Sylvia communis Whitethroat. Four eggs from Gloucestershire.
Phylloscopus trochilus Willow Warbler. Four eggs from Yorkshire.

Phylloscopus sibilatrix Wood Warbler. Two clutches (1 of 3 and 1 of 2) from Yorkshire.

Emberiza citrinella Yellowhammer. Two clutches of 4 eggs from Yorkshire.

Emberiza schoeniclus Reed Bunting. Four eggs from Norfolk.

Coccyzus americanus (Yellow-billed Cuckoo)
A clutch of 3 from Maine, North America. This is a non-parasitic cuckoo which normally builds its own nest and rears its own young. It has, occasionally, been known to lay in the nests of other birds, especially those of the same species.

STRIGIFORMES

TYTONIDAE

Tyto alba (Barn Owl)
A clutch of 4 from Yorkshire.

STRIGIDAE

Otus scops (Scops Owl)
A clutch of 5 from Asia Minor.

Bubo bubo (Eagle Owl)
A clutch of 3 from Finland.

Nyctea scandiaca (Snowy Owl)
A clutch of 5 from Lapland.

Surnia ulula (Hawk Owl)
A clutch of 7 from Lapland.

Athene noctua (Little Owl)
A clutch of 5 is devoid of data.

Strix aluco (Tawny Owl)
Two clutches (5 and 3) from Yorkshire.

Asio otus (Long-eared Owl)
Two clutches (8 and 6) from Yorkshire.

164

Asio flammeus (Short-eared Owl)
A clutch of 4 from Iceland (2nd May 1893). If the data attached to these eggs is correct, and there is no evidence to suggest otherwise, then the record predates the first recorded breeding of this species in Iceland by nearly 30 years!

Aegolius funereus (Tengmalm's Owl)
A clutch of 2 is devoid of data.

CAPRIMULGIFORMES

CAPRIMULGIDAE

Caprimulgus ruficollis (Red-necked Nightjar)
A clutch of 2 from southern Spain.

Caprimulgus europaeus (Nightjar)
Three clutches of 2 from Yorkshire.

APODIFORMES

APODIDAE

Apus apus (Swift)
Four clutches of 3 from Yorkshire.

CORACIIFORMES

ALCEDINIDAE

Alcedo atthis (Kingfisher)
A clutch of 7 from Yorkshire.

MEROPIDAE

Merops apiaster (Bee-eater)
A clutch of 5 from southern Spain.

CORACIIDAE

Coracias garrulus (Roller)
Two clutches (1 of 5 from southern Spain and 1 of 4 from Greece).

UPUPIDAE

Upupa epops (Hoopoe)
A clutch of 5 from Germany.

PICIFORMES

PICIDAE

Jynx torquilla (Wryneck)
Two clutches of 6 (Kent and Sweden)

Dendrocopus major (Great Spotted Woodpecker)
Two clutches of 5 from Yorkshire.

Picus viridis (Green Woodpecker)
A clutch of 7 from Yorkshire.

PASSERIFORMES

ALAUDIDAE

Melanocorypha leucoptera (White-winged Lark)
Three clutches of 3 from southern Russia.

Calandrella brachydactyla (Short-toed Lark)
Three clutches (1 of 5 from Belgium and 2 of 4 from Malta and southern Russia).

Galerida cristata (Crested Lark)
Three clutches (1 of 5 from Germany and 1 of 5 and 4 from southern Spain).

Lullula arborea (Woodlark)
A clutch of 4 from Suffolk.

Alauda arvensis (Skylark)
Ten clutches (1 of 5, 3 of 4, 5 of 3 and 1 of 2). All from Yorkshire with the exception of a clutch of 3 from Highland, Scotland. (See *Cuculus canorus* (Cuckoo)).

Eremophila alpestris (Shore Lark)
Two clutches of 4. One from Greenland, the other from an untraced locality.

HIRUNDINIDAE

Riparia riparia (Sand Martin)
A clutch of 5 from Yorkshire.

Hirundo rustica (Swallow)
Two clutches (5 and 3) from Yorkshire.

Delichon urbica (House Martin)
A clutch of 5 from Yorkshire.

MOTACILLIDAE

Motacilla flava (Yellow Wagtail)
Three clutches, 1 of 6 from Hereford, 1 of 5 from Sweden (said to belong to the nominate race) and 1 of 5 from Iceland. The species is but an accidental visitor to Iceland and the data attached to this clutch of eggs, assuming they are correctly identified, may be erroneous.

Motacilla cinerea (Grey Wagtail)
Five clutches (3 of 6, 1 of 5 and 1 of 4). All from Yorkshire with the exception of a clutch of 6 from Germany. (See *Cuculus canorus* (Cuckoo)).

Motacilla alba (Pied Wagtail)
Seven clutches (3 of 6, 3 of 5 and 1 of 4). Three clutches of 5 and 1 of 6 from Yorkshire are said to be of the race *yarrellii*, the clutch of 4 from Highland, Scotland is said to refer to the nominate race and 2 clutches of 6 from Iceland will be referable to this race.

Anthus campestris (Tawny Pipit)
A clutch of 5 from northern Spain.

Anthus pratensis (Meadow Pipit)
11 clutches (2 of 5, 8 of 4 and 1 of 3). All from Yorkshire with the exception of two clutches of 5 and one of 4 from Highland, Scotland. (See *Cuculus canorus* (Cuckoo)).

Anthus trivialis (Tree Pipit)
65 clutches (1of 8, 15 of 6, 37 of 5, 9 of 4 and 3 of 3) from Yorkshire. The eggs in the clutch of 8 are uniformly coloured and would appear to be the product of a single female. Clutches of this size are unusual. (See *Cuculus canorus* (Cuckoo)).

Anthus cervinus (Red-throated Pipit)
A clutch of 5 from Sweden.

Anthus petrosus (Rock Pipit)
13 clutches (6 of 5, 6 of 4 and 1 of 3). All from Yorkshire with the exception of five clutches of 4 and one of 5 from Highland, Scotland.

LANIIDAE

Lanius collurio (Red-backed Shrike)
Four clutches (1 of 7, 1 of 6 and 2 of 4). Clutches of 7, 4 and 4 from England and clutch of 6 from Austria.

Lanius minor (Lesser Grey Shrike)
Four clutches (2 of 6, 1 of 5 and 1 of 4). All from Spain with the exception of a clutch of 6 that is devoid of data.

Lanius excubitor (Great Grey Shrike)
Two clutches (5 and 6). Clutch of 6 from an untraced locality and the clutch of 5 is devoid of data.

Lanius senator (Woodchat)
Three clutches (2 of 5 and 1 of 4). Clutches of 5 and 4 from southern Spain and clutch of 5 from Portugal.

CINCLIDAE

Cinclus cinclus (Dipper)
Five clutches (4 of 5 and 1 of 4). All from Yorkshire with the exception of a clutch of 5 from Sweden.

TROGLODYTIDAE

Troglodytes troglodytes (Wren)
Five clutches (1 of 8, 3 of 7 and 1 of 4) from Yorkshire.

PRUNELLIDAE

Prunella modularis (Dunnock)
Four clutches (1 of 6, 2 of 5 and 1 of 4) from Yorkshire. (See *Cuculus canorus* (Cuckoo)).

TURDIDAE

Cercotrichas galactotes (Rufous Bushchat)
Three clutches of 5. Two from southern Spain and one from southern Germany.

Erithacus rubecula (Robin)
Five clutches of 5 from Yorkshire. (See *Cuculus canorus* (Cuckoo)).

Luscinia megarhynchos (Nightingale)
Three clutches (1 of 6, 1 of 4 and 1 of 3) from southern England. (See *Cuculus canorus* (Cuckoo)).

Luscinia svecica (Bluethroat)
Three clutches (2 of 6 and 1 of 4). The clutches of 6 (from Sweden) will belong to the nominate race whilst the clutch of 4 (from southern Germany) will belong to the race *cyanecula*.

Phoenicurus phoenicurus (Redstart)
Five clutches (1 of 7, 2 of 6, and 2 of 5) from Yorkshire.

Saxicola rubetra (Whinchat)
Four clutches (1 of 7 and 3 of 6) from Yorkshire.

Saxicola torquata (Stonechat)
Four clutches (1 of 6 and 3 of 5). Clutch of 6 from Jersey, Channel Islands and clutches of 5 from Germany, Yorkshire and Wales.

Oenanthe isabellina (Isabelline Wheatear)
A clutch of 5 from the Ukraine.

Oenanthe oenanthe (Wheatear)
Five clutches (3 of 6 and 2 of 5) from Highland, Scotland.

Oenanthe deserti (Desert Wheatear)
A clutch of 5 from Turkestan, Russia.

Oenanthe hispanica (Black-eared Wheatear)
A clutch of 5 from Spain.

Monticola saxatilis (Rock Thrush)
A clutch of 4 from an untraced locality.

Turdus torquatus (Ring Ouzel)
Eight clutches (2 of 5, 5 of 4 and 1 of 1). All from Yorkshire with the exception of the single egg from Highland, Scotland.

Turdus merula (Blackbird)
Four clutches (2 of 5 and 2 of 4) from Yorkshire.

Turdus pilaris (Fieldfare)
A clutch of 4 from Sweden.

Turdus philomelos (Song Thrush)
Five clutches (1 of 5, 2 of 4 and 2 of 3) from Yorkshire.

Turdus iliacus (Redwing)
A clutch of 6 from Iceland.

Turdus viscivorus (Mistle Thrush)
Five clutches (3 of 5 and 2 of 4) from Yorkshire. A clutch of 5 from Aislaby on 17th April 1908 contains an egg that is but half the size of the other four.

SYLVIIDAE

Locustella naevia (Grasshopper Warbler)
Five clutches (4 of 5 and 1 of 4). All from Yorkshire with the exception of the clutch of 4 from Holland. (See *Cuculus canorus* (Cuckoo)).

Acrocephalus paludicola (Aquatic Warbler)
A clutch of 5 from Germany.

Acrocephalus schoenobaenus (Sedge Warbler)
Five clutches (2 of 6, 2 of 5 and 1 of 4) from Yorkshire. (See *Cuculus canorus* (Cuckoo)).

Acrocephalus scirpaceus (Reed Warbler)
Five clutches (2 of 5 and 3 of 4). All from southern England with the exception of a clutch of 4 that is devoid of data. (See *Cuculus canorus* (Cuckoo)).

Acrocephalus palustris (Marsh Warbler)
Five clutches (4 of 5 and 1 of 4). All from Germany with the exception of two clutches of 5 that are devoid of data.

Acrocephalus arundinaceus (Great Reed Warbler)
Six clutches (4 of 5 and 2 of 4). Two clutches of 5 and one of 4 from southern Spain, clutch of 4 from Germany, whilst the other clutches are devoid of data.

Hippolais icterina (Icterine Warbler)
Two clutches of 5. One from Germany, whilst the other is devoid of data.

Sylvia nisoria (Barred Warbler)
Four clutches (2 of 5 and 2 of 4). Clutches of 5 and 4 from Poland, clutch of 5 from Germany and a clutch of 4 from France.

Sylvia hortensis (Orphean Warbler)
Four clutches (1 of 5 and 3 of 4). Clutches of 5 and 4 from southern Europe, clutch of 4 from Africa, whilst the other clutch is devoid of data.

Sylvia borin (Garden Warbler)
Five clutches of 5 from Yorkshire with the exception of one from Worcestershire.

Sylvia atricapilla (Blackcap)
Five clutches (3 of 5 and 2 of 4) from Yorkshire.

Sylvia communis (Whitethroat)
Nine clutches (2 of 6, 4 of 5, 2 of 4 and 1 of 3). All from Yorkshire with the exception of clutches of 4 and 5 from Gloucester. (See *Cuculus canorus* (Cuckoo)).

Sylvia curruca (Lesser Whitethroat)
Five clutches (1 of 6, 3 of 5 and 1 of 4) from Yorkshire.

Sylvia undata (Dartford Warbler)
A clutch of 5 from the Isle of Wight.

Phylloscopus trochilus (Willow Warbler)
15 clutches (1 of 8, 3 of 7, 7 of 6, 1 of 5 and 3 of 4). All from Yorkshire with
the exception of a clutch of 7 from Worcestershire. (See *Cuculus canorus*
(Cuckoo)).

Phylloscopus colybitus (Chiffchaff)
Five clutches (4 of 6 and 1 of 5) from Yorkshire.

Phylloscopus sibilatrix (Wood Warbler)
Nine clutches (1 of 7, 4 of 6, 3 of 5 and 1 of 4) from Yorkshire. (See *Cuculus
canorus* (Cuckoo)).

Regulus regulus (Goldcrest)
Four clutches (1 of 10, 2 of 9 and 1 of 8) from Yorkshire.

Regulus ignicapillus (Firecrest)
A clutch of 9 from Germany.

MUSCICAPIDAE

Ficedula hypoleuca (Pied Flycatcher)
Five clutches (1 of 10, 1 of 6 and 3 of 5) from Yorkshire. This species normally lays 6 or 7 eggs, sometimes only 4 or 5, and the clutch of 10 may
be the product of two females.

Ficedula parva (Red-breasted Flycatcher)
A clutch of 2, still intact in a nest, is devoid of data.

Muscicapa striata (Spotted Flycatcher)
Seven clutches (6 of 5 and 1 of 4) from Yorkshire.

TIMALIIDAE

Panurus biarmicus (Bearded Tit)
A clutch of 5 from Great Yarmouth, Norfolk.

AEGITHALIDAE

Aegithalos caudatus (Long-tailed Tit)
Four clutches (2 of 12, 1 of 10 and 1 of 9) from Yorkshire.

PARIDAE

Parus palustris (Marsh Tit)
Five clutches (1 of 9, 1 of 7, 1 of 6 and 2 of 5). All from Yorkshire with the exception of the clutch of 6 from Dyfed, Wales. These eggs were collected before it had been realised that Willow Tit *P. montanus* was masquerading under this species' name and it is possible that some of these clutches should be referred to that species.

Parus ater (Coal Tit)
Four clutches (1 of 10, 2 of 8 and 1 of 7). All from Yorkshire with the exception of a clutch of 8 from Gloucestershire.

Parus cristatus (Crested Tit)
Three clutches (1 of 8 and 2 of 5) from Sweden.

Parus major (Great Tit)
11 clutches (4 of 10, 2 of 9, 3 of 8 and 2 of 7) from Yorkshire.

Parus caeruleus (Blue Tit)
Seven clutches (2 of 13, 1 of 11, 1 of 10, 2 of 9 and 1 of 8) from Yorkshire.

SITTIDAE

Sitta europaea (Nuthatch)
Four clutches (1 of 9, 1 of 6, 1 of 5 and 1 of 3) from southern England.

CERTHIIDAE

Certhia familiaris (Treecreeper)
Six clutches (2 of 6, 3 of 5 and 1 of 4) from Yorkshire.

EMBERIZIDAE

Miliaria calandra (Corn Bunting)
Seven clutches (3 of 5, 3 of 4 and 1 of 2). All from Muston, Yorkshire with the exception of a clutch of 4 from Highland, Scotland.

Emberiza citrinella (Yellowhammer)
Eight clutches (1 of 5, 5 of 4 and 2 of 3) from Yorkshire. (See *Cuculus canorus* (Cuckoo)).

Emberiza hortulana (Ortolan Bunting)
Two clutches of 4. One from Asia Minor, the other from Portugal.

Emberiza cirlus (Cirl Bunting)
Three clutches (1of 5, 1 of 4 and 1 of 3). Clutch of 3 from Greece, the others from southern England.

Emberiza melanocephala (Black-headed Bunting)
Two clutches (1 of 6 from southern Russia and 1 of 4 from Yugoslavia).

Emberiza schoeniclus (Reed Bunting)
Seven clutches (5 of 5 and 2 of 4). All from Yorkshire with the exception of a clutch of 5 from Sussex and one of 4 from Highland, Scotland. (See *Cuculus canorus* (Cuckoo)).

Calcarius lapponicus (Lapland Bunting)
Four clutches (1 of 7, 1 of 6 and 2 of 5). All from Lapland (clutch of 7 from Swedish Lapland) with the exception of a clutch of 5 from Greenland.

Plectrophenax nivalis (Snow Bunting)
Three clutches of 5 from Iceland.

FRINGILLIDAE

Fringilla coelebs (Chaffinch)
Seven clutches (1 of 6, 4 of 5 and 2 of 4). All from Yorkshire with the exception of a clutch of 5 from Devon and clutches of 4 from Wiltshire and Wales.

Fringilla montifringilla (Brambling)
Four clutches (1 of 7, 2 of 6 and 1 of 5). Clutches of 7 and 6 from Norway, a clutch of 6 from Lapland and a clutch of 5 from Sweden.

Serinus serinus (Serin)
A clutch of 4 from Portugal.

Carduelis chloris (Greenfinch)
Six clutches (1 of 7, 4 of 5 and 1 of 3) from Yorkshire.

Carduelis spinus (Siskin)
A clutch of 6 from Germany.

Carduelis carduelis (Goldfinch)
Five clutches (1 of 6, 3 of 5 and 1 of 4) from Yorkshire.

Carduelis flammea (Redpoll)
11 clutches (1 of 6, 7 of 5, 2 of 4 and 1 of 3). All from Yorkshire with the exception of clutches of 4 and 3 from Iceland. These last mentioned clutches will be referable to the race *islandica*.

Carduelis flavirostris (Twite)
Four clutches (1 of 7, 1 of 6 and 2 of 5). All from Steeton Moor, Yorkshire with the exception of a clutch of 5 from Orkney, Scotland.

Carduelis cannabina (Linnet)
Six clutches (1 of 6, 2 of 5 and 3 of 4) from Yorkshire.

Pinicola enucleator (Pine Grosbeak)
A clutch of 4 from Sweden.

Loxia curvirostra (Crossbill)
A clutch of 4 from Sweden.

Pyrrhula pyrrhula (Bullfinch)
Six clutches of 5. All from Yorkshire with the exception of a clutch from Norfolk.

Coccothraustes coccothraustes (Hawfinch)
Two clutches (1 of 5 and 1 of 4) from southern England.

PASSERIDAE

Passer domesticus (House Sparrow)
Six clutches (4 of 5 and 2 of 4) from Yorkshire.

Passer montanus (Tree Sparrow)
Seven clutches (1 of 6, 4 of 5 and 2 of 4) from Yorkshire.

STURNIDAE

Sturnus vulgaris (Starling)
Six clutches (3 of 6, 1 of 5 and 1 of 3) from Yorkshire.

Sturnus roseus (Rose-coloured Starling)
A clutch of 4 from southern Russia.

ORIOLIDAE

Oriolus oriolus (Golden Oriole)
Two clutches (1 of 5 from Spain and 1 of 4 from Germany).

CORVIDAE

Garrulus glandarius (Jay)
Five clutches (3 of 6 and 2 of 4) from Yorkshire.

Pica pica (Magpie)
Eight clutches (7 of 7 and 1 of 6) from Yorkshire.

Nucrifraga caryocatactes (Nutcracker)
Two clutches of 4 (one from Switzerland, the other from an untraced locality).

Pyrrhocorax pyrrhocorax (Chough)
Two clutches of 4 from north Wales.

Corvus monedula (Jackdaw)
Eight clutches (5 of 5, 2 of 4 and 1 of 2) from Yorkshire.

Corvus frugilegus (Rook)
Three clutches (1 of 4, 1 of 3 and 1 of 2) from Yorkshire.

Corvus corone (Carrion Crow)
12 clutches (1 of 7, 8 of 5 and 3 of 4). All from Yorkshire with the exception of three clutches of 5 and one of 4 from Scotland. The eggs from Yorkshire belong to the nominate race whilst the eggs from Scotland are said to belong to the race *cornix*.

Corvus corax (Raven)
Two clutches (1 of 4 and 1 of 2) from northern Scotland.

ADDITIONAL EGGS

STRUTHIONIFORMES

STRUTHIONIDAE

176

Struthio camelus (Ostrich)
Five eggs, one of which originates from South Africa.

DROMAIIDAE

Dromaius novaehollandiae (Emu)
A single egg from Australia.

SKELETONS

The Museum possesses a total of 39 named bird skeletons (including six only named to genus). As with the eggs it has not been possible to apply a critical standard of identification. As no geographical area of acquisition is indicated, the identification of these specimens is extremely problematical. It must also be remembered that the person involved with identification in the first instance was in the privileged position of having plumage details with which to work! With few exceptions the identification of these specimens must, therefore, remain with the initial determiner. Other, unnamed skeletons remain for the dedicated specialist to identify. For completeness, and as a service to ornithologists, the names appended to these skeletal remains are listed below.

STRUTHIONIFORMES

STRUTHIONIDAE

Struthio camelus (Ostrich)

DINORNITHIFORMES

DINORNITHIIDAE

Dinornis robustus (Moa)
At the time of acquisition (1876), this skeleton was described as the largest and most perfect specimen known to science.

CASUARIIFORMES

CASUARIIDAE

Casuarius sp. (Cassowary sp.)

APTERYGIFORMES

APTERYGIDAE

178

Apteryx haastii (Great Spotted Kiwi)

SPHENISCIFORMES

SPHENISCIDAE

Spheniscus demersus (Jackass Penguin)

GAVIIFORMES

GAVIIDAE

Gavia stellata (Red-throated Diver)
Although said to be this species, the head and bill are not those of *G. stellata*.

Gavia arctica (Black-throated Diver)
Two specimens. One of the specimens is without head and bill.

PODICIPEDIFORMES

PODICIPEDIDAE

Tachybaptus ruficollis (Little Grebe)

PELECANIFORMES

PELECANIDAE
Pelecanus sp. (Pelican sp.)

SULIDAE

Morus bassanus (Gannet)
Unfortunately no head and bill are attached.

CICONIIFORMES

ARDEIDAE

Ardea purpurea (Purple Heron)

Ardea cinerea (Grey Heron)

CICONIIDAE

Ciconia ciconia (White Stork)

Leptoptilus dubius (Greater Adjutant)

Leptoptilus crumeniferus (Marabou Stork)

THRESKIORNITHIDAE

Platalea leucorodia (Spoonbill)

ANSERIFORMES

ANATIDAE

Cygnus cygnus (Whooper Swan)

Cygnus columbianus (Bewick's Swan)

Cygnus sp. (Swan sp.)
Two specimens.

Anser anser (Greylag Goose)
Unfortunately no head and bill are attached.

Anas platyrhynchos (Mallard)
A specimen labelled "Wild Duck, *Anas boschas,* var. Penquin Duck" is presumably referable to the race of Mallard found in Greenland *A. p. conboschas.* Unfortunately no head and bill are attached.

FALCONIFORMES

CATHARTIDAE

Sarcorhamphus papa (King Vulture)

ACCIPITRIDAE

Haliaeetus albicilla (White-tailed Sea Eagle)

Gypohierax angolensis (Palm-nut Vulture)

Gyps fulvus (Griffon Vulture)

Circus aeruginosus (Marsh Harrier)

SAGITTARIIDAE

Sagittarius serpentarius (Secretary Bird)

GALLIFORMES

CRACIDAE

Penelope sp. (Guan sp.)
The specimen is labelled *"Penelope cristata"*, which must be an old, unfound, synonym.

Crax rubra (Great Curassow)

MELEAGRIDAE

Meleagris gallopavo (Turkey)

TETRAONIDAE

Tetrao urogallus (Capercaillie)

PHASIANIDAE

Pavo cristatus (Peafowl)
Only the lower mandible remains of the head and bill.

GRUIFORMES

GRUIDAE

Grus sp. (Crane sp.)
The specimen is labelled *"Crus cinerea"*, which must be an old, unfound, synonym.

CHARADRIIFORMES

SCOLOPACIDAE

Numenius arquata (Curlew)

LARIDAE

Larus marinus (Great Black-backed Gull)

ALCIDAE

Pinguinus impennis (Great Auk)
Eight bones from Funk Island, off the New Foundland coast, in 1880 have been labelled: femur, tibia, pelvis and lower mandible.

COLUMBIFORMES

COLUMBIDAE

Raphus cucullatus (Dodo)
23 bones are boxed together and have been labelled: sternum, upper jaw, lower jaw, scapular and coracoid, femur, tibia and fibula, tarso-metatarsus and humerus. Two fossil bones in a case were acquired on Mauritius, Indian Ocean (where the species was endemic).

STRIGIFORMES

STRIGIDAE

Asio otus (Long-eared Owl)
Unfortunately no head and bill are attached.

PASSERIFORMES

CORVIDAE

Corvus corone (Carrion Crow)
Two specimens. One of the specimens is labelled "Carrion Crow, *Corvus corone*" the other "Hooded Crow, *Corvus cornix*".

RESOURCES AT THE YORKSHIRE MUSEUM

The various bird collections held at the Yorkshire Museum are open for consultation by anyone with an interest in ornithology. People wishing to inspect these collections are invited to contact the Yorkshire Museum at the address below.

The Museum can make available a work area with a full range of measuring equipment. Important identification publications include the full complement of *The Birds of the Western Palearctic* along with Kevin Baker's *Identification Guide to European non-passerines* and the most recent (4th edition) *Identification Guide to European Passerines* by Lars Svennson.

The Museum's extensive library also holds a number of important historical publications i.e. Morris's *A Natural History of the Nests and Eggs of British Birds* (1875), Bree's *A History of the Birds of Europe, not observed in the British Isles* (1875), Seebohm's *A History of British Birds, with coloured illustrations of their Eggs* (1883), Nelson's *The Birds of Yorkshire* (1907) and Thorburn's *British Birds* (1915).

Yorkshire Museum, Museum Gardens
York YO1 2DR, United Kingdom
Telephone: YORK (01904) 629745   Fax: YORK (01904) 651221

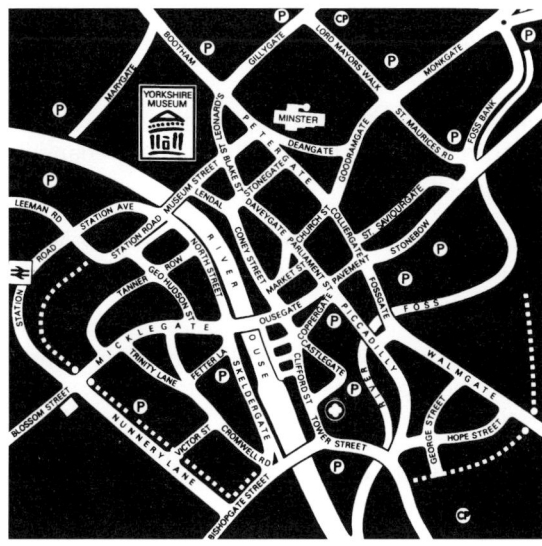

ACKNOWLEDGEMENTS

The compilation of *Birds in the Yorkshire Museum* would have been a far more daunting undertaking were it not for the computerisation of the relevant data. This computerisation ran in parallel with the curation of the collections and I will forever be indebted to Pamela Welsh for this service. Pam worked at the Museum on a short term contract and spent most of her time inputting this essential information.

The Museum's Registrar, Melanie Baldwin, was instrumental in supplying the necessary information from this data base. Melanie also did an admirable job in shouldering the responsibility for the numerous computerisation problems that ensued.

Most of the information in the section on *History* was gleaned from the *Yorkshire Philosophical Society Annual Reports* and the Museum's Curatorial Assistant, Janetta Lambert, must be thanked for playing a vital role in the extraction of this information.

John R. Mather very kindly undertook the responsibility of checking my attempts at assessing breeding distribution for each species and I thank him wholeheartedly for his advice.

For promptly supplying ringing information relating to a ringed Sparrowhawk I would like to thank Jackie Clarke of the Ringing Unit at the British Trust for Ornithology.

The identification of certain wildfowl ducklings and a gosling was carried out by Martin Brown of the Wildfowl and Wetland Trust and his efforts for this service are to be applauded.

Michael Walters, egg curator at the Natural History Museum (Tring), deserves my thanks for giving much advice on the problematical bustard eggs.

Peter Colston of the Natural History Museum, Tring, and Per Alstrom are to be thanked for expending time and effort on the determination of a problematical *Phylloscopus* warbler.

Thanks must also go to the County Museums Officer, Brian Hayton, for his advice and encouragement throughout this project.

INDEX OF SCIENTIFIC NAMES

BS = Birds skins   MC = Mounts (cased)   MM = Mounts (multi-species cased)   MU = Mounts (uncased)   EM = Eggs (Museum coll.)   EC = Eggs (W. Cooper coll.)   SK = Skeletons

	BS	MC	MM	MU	EM	EC	SK
Accipiter gentilis	21			114		153	
Accipiter nisus	21			114	134	154	
Aceros undulatus				125			
Acridotheres tristis	85						
Acrocephalus arundinaceus	66					171	
Acrocephalus bistrigiceps	66						
Acrocephalus dumetorum	66						
Acrocephalus melanopogon	66						
Acrocephalus paludicola	65					170	
Acrocephalus palustris	66					171	
Acrocephalus schoenobaenus	65				143	170	
Acrocephalus scirpaceus	66				143	171	
Actitis hypoleucos	32			119	136	159	
Actitis macularia	32						
Aegithalos caudatus	72			128	144	172	
Aegolius funereus	45					165	
Alauda arvensis	50	101			140	166	
Alca torda	40	99		121	138	162	
Alcedo atthis	46	101			140	165	
Alectoris barbara	25						
Alectoris chukar	25						
Alectoris rufa	25			116	134	155	
Alle alle	40	98	106	121		162	
Alopochen aegytiacus	15						
Amandava subflava	83						
Amazona amazonica				123			
Anas acuta	16			112		151	
Anas americana	16					151	
Anas discors						151	
Anas clypeata	16			112		151	
Anas crecca	16			112	133	151	
Anas formosa	16						
Anas penelope	15	92		112		151	

	BS	MC	MM	MU	EM	EC	SK
Anas platyrhynchos	16			112	133	151	180
Anas platyrhynchos X A. acuta				112			
Anas querquedula	16					151	
Anas strepera						151	
Anous stolidus						162	
Anser albifrons				111		150	
Anser anser	15			111		150	180
Anser brachyrhynchus				111		150	
Anser erythropus	15						
Anser fabalis						150	
Anthropoides virgo	26					155	
Anthus berthelotii	54						
Anthus campestris	53					167	
Anthus cervinus	53					168	
Anthus hodgsoni	53						
Anthus petrosus	54				141	168	
Anthus pratensis	53	102			141	167	
Anthus richardi	53						
Anthus roseatus	53						
Anthus rubescens	54						
Anthus spinoletta	54						
Anthus trivialis	53				141	168	
Apteryx australis			106	108			
Apteryx haastii				108			179
Apteryx owenii /A. haastii			106	108			
Apus affinis	46						
Apus apus	45	100			139	165	
Apus melba	45	100					
Apus pallidus	45						
Apus unicolor	45						
Aquila chrysaetos				115		154	
Aquila clanga	22					154	
Ardea cinerea	14			110	133	149	179
Ardea purpurea	14			110		149	179
Ardoela ralloides	13	91				149	
Arenaria interpres	32		105	119	136	159	
Asio capensis	45						
Asio flammeus	44	100		125		165	
Asio otus	44			125	139	164	182
Athene noctua	44				139	164	

	BS	MC	MM	MU	EM	EC	SK
Aythya ferina	17			112		152	
Aythya fuligula	17			113		152	
Aythya marila.	17	92		113		152	
Aythya nyroca	17					152	
Aythya valisineria	17						
Bartramia longicauda						158	
Bombycilla garrulus	57						
Bombycilla japonica	57						
Bonasa bosania	25						
Botaurus lentiginosus						149	
Botaurus stellaris	13	91			133	149	
Branta bernicla	15			111			
Branta canadensis		92					
Branta leucopsis				111			
Bubo bubo	44	100		124		164	
Bubulcus ibis						149	
Bucanetes githaginea	80						
Bucephala clangula	18	92				152	
Buceros bicornis				125			
Buceros rhinoceros				125			
Bulweria bulwerii		90		109		148	
Burhinus oedicnemus	29	95		118	135	157	
Buteo buteo	21	93			134	154	
Buteo lagopus	21	93		115		154	
Buteo rufinus	21						
Butorides virescens						149	
Calandrella brachydactyla	50					166	
Calandrella rufescens	50						
Calcarius lapponicus	77					174	
Calidris alba	34	97	105				
Calidris alpina	34	97	105	120	137	159	
Calidris bairdii	34						
Calidris canutus	33		105	119			
Calidris ferruginea	35	97					
Calidns maritima	34			119		159	
Calidris melanotos	34	97					
Calidris minuta	34						
Calidris temminckii	34	97			136	159	
Calyptomena viridis				126			

	BS	MC	MM	MU	EM	EC	SK
Caprimulgus europaeus	45			125	139	165	
Caprimulgus rufcollis						165	
Carduelis cannabina	80				145	175	
Carduelis carduelis	79	103	106	130		175	
Carduelis carduelis X							
Serinus canaria	79						
Carduelis chloris	78		106		145	174	
Carduelis flammea	79			130	145	175	
Carduelis flavirostris	80					175	
Carduelis hornemanni	79						
Carduelis sinica	78						
Carduelis spinus	79			129		174	
Carpodacus erythrinus	81						
Carpodacus roseus	81						
Carpodacus vinaceus	81						
Casuarius sp							178
Cepphus grylle	41	99		122		162	
Cercotrichas galactotes	58					169	
Certhia familiaris	74			129	144	173	
Cettia cetti	65						
Chalcophaps indica			107				
Charadrius alexandrinus	30		105			158	
Charadrius asiaticus	30						
Charadrius dubius	30					158	
Charadrius hiaticula	30		105		136	158	
Charadrius morinellus	30	96				158	
Charadrius vociferus						158	
Charmosyna papou				123			
Chlamydotis undulata		95				157	
Chlidonias hybrida						161	
Chlidonias leucoptera	38					161	
Chlidonias nigra	38	98				161	
Chloephaga picta				111			
Chrysococcyx lucidus			107				
Chrysolophus pictus	26			117			
Ciconia ciconia						150	180
Ciconia nigra		91				150	
Cinclus cinclus	57			126	141	168	
Circaetus gallicus	20						
Circus aeruginosus	20			114	133	153	181
Circus cyaneus	20			114		153	

188

	BS	MC	MM	MU	EM	EC	SK
Circus macrourus	20						
Circus pygargus	21			134	153		
Cisticola juncidis	71						
Clamator glandarius						163	
Clangula hyemalis	18	92				152	
Coccyzus americanus	43					164	
Coccothraustes coccothraustes	83			130	145	175	
Columba livia	41			122	138	163	
Columba oenas	42			122	138	163	
Columba palumbus	42			122	138	163	
Coracias garrulus	47					165	
Coracina lineata			107				
Cormobates leucophaea			107				
Corvus corax	88			131		176	
Corvus corone	87	104		131	146	176	182
Corvus corone X C. corax	87						
Corvus frugilegus	87	104		131	146	176	
Corvus monedula	87	103		131	146	176	
Corvus torquatus	88						
Coturnix coturnix	25					155	
Coturnix japonica	25						
Cracticus torquatus			107				
Crax rubra							181
Crex crex	27			117	135	156	
Cuculus canorus	43			124	138	163	
Cuculus pyrrhophanus			107				
Cursorius cursor						157	
Cyanopica cyana	86						
Cygnus columbianus				111			180
Cygnus cygnus				111		150	180
Cygnus olor	15			111		150	
Cygnus sp							180
Delichon urbica	51	101			140	167	
Dendrocopus leucotos	47						
Dendrocopus major	48			126		166	
Dendrocopus medius	48						
Dendrocopus minor	47			126			
Dendrocopus syriacus	48						
Dendrocygna javanica	15						
Dicaeum hirundinaceum			107				

	BS	MC	MM	MU	EM	EC	SK
Dicrurus paradiseus				130			
Dinornis robustus							178
Dromaius novaehollandiae						177	
Dryocopus pileatus	48						
Eclectus roratus				123			
Ectopistes migratorius	42	99					
Egretta alba	14	91				149	
Egretta garzetta	13					149	
Elanus caeruleus	19						
Emberiza aureola	76						
Emberiza bruniceps	76						
Emberiza caesia	75						
Emberiza cia	75						
Emberiza cioides	75						
Emberiza cirlus	76					174	
Emberiza citrinella	74		106		145	173	
Emberiza citrinella X E. leucocephala	75						
Emberiza fucata	76						
Emberiza hortulana	75					174	
Emberiza leucocephala	75						
Emberiza melanocephala	76					174	
Emberiza pusilla	76						
Emberiza rustica	76	103					
Emberiza schoeniclus	77	103	106		145	174	
Emberiza spodocephala	77						
Emberiza sulphurata	77						
Eopsaltria australis			107				
Eos reticulata				123			
Eremophila alpestris	51					167	
Erithacus rubecula	58	102	106	127	142	169	
Erythrura gouldiae			107				
Eudynamys scolopacea			107				
Eurynorhynchus pygmeus	35						
Eurystomus orientalis			107				
Falco cherrug				116			
Falco columbarius	23			115	134	154	
Falco eleonorae	23						
Falco naumanni				115		154	

	BS	MC	MM	MU	EM	EC	SK
Falco peregrinus	24		105	116		155	
Falco rusticolus	23			115		154	
Falco sparverius	22						
Falco subbuteo	23	93		115	134	154	
Falco tinnunculus	23	93		115	134	154	
Falco vespertinus	23	93				154	
Ficedula albicollis	71						
Ficedula hypoleuca	71			128		172	
Ficedula parva	71					172	
Ficedula zanthopygia	71						
Fratercula arctica	41	99			138	162	
Fringilla coelebs	77		106	129	145	174	
Fringilla montifringilla	78			129		174	
Fringilla teydea	77						
Fulica atra	28			118	135	156	
Fulica cristata	28						
Fulmarus glacialis	10	90		109	132	147	
Galerida cristata	50					166	
Gallinago gallinago	33	97			136	159	
Gallinago media	33	97					
Gallinago stenura	33						
Gallinula chloropus	27			117	135	156	
Galliralus australis	26						
Galloperdix sp		94					
Garrulus glandarius	86	103		130	146	176	
Gavia arctica	9			109		147	179
Gavia immer	9			109		147	
Gavia stellata	9			108		147	179
Gelochelidon nilotica	39					161	
Glareola pratincola	29			118		157	
Goura cristata				122			
Grus grus		94				155	
Grus sp							181
Gypohierax angolensis							181
Gyps fulvus						153	181
Haematopus ostralegus	28			118	135	157	
Halcyon macleayii			107				
Halcyon pileata	46						
Haliaeetus albicilla	20			114		153	180

	BS	MC	MM	MU	EM	EC	SK
Hieraaetus fasciatus	22						
Hieraaetus pennatus	22						
Himantopus himantopus						157	
Hippolais icterina	66					171	
Hippolais olivetorum	67						
Hippolais pallida	67						
Hippolais polyglotta	66						
Hirundo daurica	51						
Hirundo rustica	51	101	106		140	167	
Histrionicus histrionicus	18					152	
Hydrobates pelagicus	12	90				148	
Irania guttularis	59						
Ixobrychus minutus	13						
Jynx torquilla	47			126	140	166	
Lagopus lagopus	24	93		116	134	155	
Lagopus mutus	24		105	116		155	
Lanius bucephalus	55						
Lanius collurio	55				141	168	
Lanius cristatus	55						
Lanius excubitor	56					168	
Lanius isabellinus	55						
Lanius ludovicianus	56						
Lanius minor	56					168	
Lanius nubicus	57						
Lanius schach	55						
Lanius senator	56					168	
Lanius sphenocercus	56						
Lanius tigrinus	55						
Larus argentatus	36	98	106		137	160	
Larus cachinnans	36						
Larus canus	36		106	120	137	160	
Larus fuscus	36			120	137	160	
Larus genei	38						
Larus glaucoides	37			121		160	
Larus hyperboreus	37			121		160	
Larus ichthyaetus						160	
Larus marinus	37			121		160	182
Larus melanocephalus	37						

	BS	MC	MM	MU	EM	EC	SK
Larus minutus	38	98				161	
Larus pacificus	36						
Larus philadelphia	38					161	
Larus pipixcan	37						
Larus ridibundus	37	98		121	137	161	
Leptoptilus crumeniferus							180
Leptoptilus dubius							180
Leucosticte arctoa	80						
Leucosticte brandti	80						
Leucosticte nemoricola	80						
Limicola falcinellus						160	
Limosa lapponica	30		105	119		158	
Limosa limosa	30			119		158	
Locustella certhiola	65						
Locustella fluviatilis	65						
Locustella lanceolata	65						
Locustella naevia	65				143	170	
Loxia curvirostra	82			130		175	
Loxia leucoptera	82						
Loxia pytyopsittacus	81						
Loxia scotica	82						
Lullula arborea	50	101			140	166	
Luscinia calliope	59						
Luscinia luscinia	58						
Luscinia megarhynchos	59	102			142	169	
Luscinia svecica	59	102				169	
Lymnocryptes minima	33	97				159	
Marmaronetta angustirostris	16						
Melanitta fusca	18			113		152	
Melanitta nigra	18			113		152	
Melanocorypha calandra	49						
Melanocorypha leucoptera	49					166	
Melanocorypha yeltoniensis	50						
Meleagris gallopava							181
Mergus albellus				113			
Mergus cucullatus				113		152	
Mergus merganser	19	92	105	113	133	153	
Mergus serrator	19			113		152	
Merops apiaster	47					165	
Merops orientalis	46						

	BS	MC	MM	MU	EM	EC	SK
Merops superciliosus	46						
Miliaria calandra	74			129	144	173	
Milvus migrans	19					153	
Milvus milvus	20	93		114		153	
Monticola saxatilis	63					170	
Monticola solitarius	63						
Montifringilla nivalis	85						
Morus bassanus	12			110		148	179
Motacilla alba	52				141	167	
Motacilla cinerea	52	102			141	167	
Motacilla citreola	52						
Motacilla flava	52				141	167	
Motacilla flaviventris	52						
Motacilla maderaspatensis	52						
Muscicapa striata	72				144	172	
Neophema bourkii	43						
Neophron percnopterus						153	
Neffa rufina	17			112		151	
Nucrifraga caryocatactes	87					176	
Numenius arquata	31	96			136	158	182
Numenius phaeopus	31				136	158	
Nyctea scandiaca				124		164	
Nycticorax nycticorax	13	91				149	
Oceanites oceanicus	11						
Oceanodroma leucorhoa	12	90				148	
Oenanthe deserti	61					169	
Oenanthe finschii	61						
Oenanthe hispanica	61					170	
Oenanthe isabellina	61					169	
Oenanthe leucopyga	62						
Oenanthe leucura	62						
Oenanthe lugens	62						
Oenanthe moesta	62						
Oenanthe oenanthe	61	102			142	169	
Oenanthe picata	62						
Oenanthe pleschanka	62						
Oriolus oriolus	85	103				176	
Otis tarda	28	95		118	135	156	
Otus scops	44	100				164	

	BS	MC	MM	MU	EM	EC	SK
Pachycephala pectoralis			107				
Pagophiia eburnea	36						
Pandion haliaetus	19	92		114	133	153	
Panurus biarmicus						172	
Paradisaea raggiana	86						
Pardalotus punctatus			107				
Pardalotus striatus			107				
Parus ater	73				144	173	
Parus atricapillus	73						
Parus caeruleus	73		106	129	144	173	
Parus cristatus	73					173	
Parus lugubris	72						
Parus major	73			129	144	173	
Parus montanus	73				144		
Parus palustris	72			129	144	173	
Passer ammondendri	84						
Passer domesticus	83		106	130	145	175	
Passer domesticus /							
P. hispaniolensis	84						
Passer hispaniolensis	83						
Passer moabiticus	84						
Passer montanus	84		106	130	146	175	
Passer rutilans	84						
Pava cristatus				117			181
Pelagodroma marina					148		
Pelecanus sp							179
Penelope sp							181
Perdix perdix	25		106		134		
Pernis apivorus	19			114		153	
Petroica rosea			107				
Petronia petronia	84						
Phalacrocorax aristotelis	12	91		110	133	149	
Phalacrocorax carbo	12			110	132	148	
Phalacrocorax pygmeus	13						
Phalaropus fulicarius	32	96				159	
Phalaropus lobatus	32	96				159	
Phasianus colchicus	26	94	105	117			
Philomacus pugnax	35	97	105	120	137	160	
Phoenicopterus ruber						150	
Phoenicurus auroreus	60						
Phoenicurus fuliginosus	60						

	BS	MC	MM	MU	EM	EC	SK
Phoenicurus moussieri	60						
Phoenicurus ochruros	59	102					
Phoenicurus phoenicurus	60			127	142	169	
Phylloscopus affinis	69						
Phylloscopus borealis	70						
Phylloscopus collybita	69				143	172	
Phylloscopus fuscatus	69						
Phylloscopus inornatus	70						
Phytloscopus sibilatrix	69			128	144	172	
Phylloscopus trochilus	69			128	143	172	
Phylloscopus sp	70						
Pica pica	86	103		131	146	176	
Picoides tridactylus	48						
Picus awokera	49						
Picus canus	49						
Picus viridis	49	101		126		166	
Pinguinus impennis				121			182
Pinicola enucleator	81			130		175	
Platalea flavipes	15						
Platalea leucorodia	14	92				150	180
Plectrophenax nivalis	77	103				174	
Plegadis falcinellus	14					150	
Pluvialis apricaria	29	96			135	158	
Pluvialis fulva	29						
Pluvialis squatarola	29	96					
Pluvianus aegyptius	29						
Podiceps auritus	10		105	109		147	
Podiceps cristatus	10			109	132	147	
Podiceps grisegena	9			109		147	
Podiceps nigricollis	10					147	
Porphyrio porphyrio	27	95				156	
Porzana carolina	27						
Porzana parva	27					156	
Porzana porzana	27			117		156	
Porzana pusilla	27	94				156	
Prunella collaris	58						
Prunella modularis	58		106	127	142	169	
Psittacus erithacus				123			
Pterocles alchata	41						
Pterocles orientalis	41						
Ptiloris victoriae	86						

	BS	MC	MM	MU	EM	EC	SK
Ptyonoprogne rupestris	51						
Puffinus assimilis	11					148	
Puffinus gravis	10					148	
Puffinus griseus	10			110		148	
Puffinus mauretanicus		90					
Puffinus puffinus	11	90		110		148	
Pycnonotus barbatus	54						
Pycnonotus xanthopygos	54						
Pyrrhocorax graculus	87						
Pyrrhocorax pyrrhocorax	87	103		131	146	176	
Pyrrhula aurantiaca	82						
Pyrrhula erythaca	82						
Pyrrhula pyrrhula	83		106		145	175	
Rallus aquaticus	26	94		117		155	
Raphus cucullatus							182
Recurvirostra avosetta	28			118		157	
Regulus calendula	70						
Regulus ignicapillus	71					172	
Regulus regulus	71		105	128		172	
Regulus satrapa	70						
Remiz pendulinus	72						
Rhamphastos toco				126			
Rhodopechys obsoleta	80						
Rhodostethia rosea	38						
Riparia riparia	51	101	106		140	167	
Rissa tridactyla	38		106	121	137	161	
Sagittarius serpentarius							181
Sarcorhamphus papa							180
Saxicola caprata	61						
Saxicola leucura	60						
Saxicola rubetra	60			127	142	169	
Saxicola torquata	60			127	142	169	
Scolopax rusticola	33			119	136	159	
Seicercus burkii	70						
Seicercus xanthoschista	70						
Sericulus chrysocephalus			107				
Serinus canaria	78						
Serinus citrinella	78						
Serinus pusillus	78						

	BS	MC	MM	MU	EM	EC	SK
Serinus serinus	78					174	
Sitta europaea	73			129		173	
Sitta kruperi	74						
Sitta neumayer	74						
Somateria fischeri	18						
Somateria mollissima	17		106	113	133	152	
Somateria spectabilis	17					152	
Spheniscus demersus							179
Stercorarius longicaudus	36			120		160	
Stercorarius parasiticus	35		106	120		160	
Stercorarius pomarinus	35			120			
Stercorarius skua	35					160	
Sterna albifrons	39	98			137	162	
Sterna bengalensis	40						
Sterna caspia	39					161	
Sterna dougllii					137	161	
Sterna fuscata						162	
Sterna hirundo	39		106		137	161	
Sterna hirundo / S. paradisaea					137		
Sterna maximus	40						
Sterna paradisaea	39				137	161	
Sterna sandvicensis	40				138	162	
Streptopelia chinensis	43						
Streptopelia orientalis	42						
Streptopelia turtur	42			122	138		
Strigops habroptilus			106				
Strix aluco	44	100		124	139	164	
Struthio camelus				108		177	178
Sturnus roseus	85					175	
Sturnus unicolor	85						
Sturnus vulgaris	85		106	130	146	175	
Surnia ulula	44					164	
Sylvia atricapilla	67			128	143	171	
Sylvia borin	67			128	143	171	
Sylvia cantillans	68						
Sylvia communis	67			128	143	171	
Sylvia curruca	68				143	171	
Sylvia hortensis	67					171	
Sylvia melanocephala	68						
Sylvia melanothorax	68						
Sylvia nana	68						

	BS	MC	MM	MU	EM	EC	SK
Sylia nisoria	67					171	
Sylvia ruppelli	68						
Sylvia undata	69				143	171	
Syrmaticus reevesii		94	105				
Syrrhaptes paradoxus	41	99				163	
Tachybaptus ruficollis	9	89	105		132	147	179
Tadorna ferruginea						151	
Tadorna tadorna	15			112		151	
Tarsiger cyanurus	59						
Tchagra senegala	54						
Tetrao tetrix	24			116		155	
Tetrao tetrix X T. urogallus				105			
Tetrao urogallus	24	94	105	116		155	181
Tetrax tetrax	28	95				156	
Tichadroma muraria	74						
Trichoglossus haematodus				123			
Tringa erythropus	31			119			
Tringa flavipes		96					
Tringa glareola	31	96				159	
Tringa nebularia	31					159	
Tringa ochropus	31	96		119			
Tringa stagnatalis	31						
Tringa totanus	31	96			136	159	
Troglodytes troglodytes	58			127	141	168	
Turdus iliacus	64			128		170	
Turdus merula	63			127	142	170	
Turdus naumanni	64						
Turdus obscurus	64						
Turdus philomelos	64			128	142	170	
Turdus pilaris	64					170	
Turdus ruficollis	64						
Turdus torquatus	63				142	170	
Turdus viscivorus	64				143	170	
Turnix sylvatica	26						
Turnix varia			107				
Tyto alba	43	100		124	139	164	
Upupa epops	47			125	140	166	
Uragus sibiricus	81						
Uria aalge	40	99		122	138	162	

	BS	MC	MM	MU	EM	EC	SK
Uria lomvia						162	
Vanellus gregarius						158	
Vanellus vanellus	29			118	135	157	
Xema sabini	38	98					
Xenus cinereus	32						
Zoothera dauma		102		127			
Zoothera lunulata	63						

INDEX OF ENGLISH NAMES

BS = Birds skins   MC = Mounts (cased)   MM = Mounts (multi-species cased)   MU = Mounts (uncased)   EM = Eggs (Museum coll.)   EC = Eggs (W. Cooper coll.)   SK = Skeletons

	BS	MC	MM	MU	EM	EC	SK
Bunting, Cretzschmar's	75						
Bunting, Japanese Yellow	77						
Bunting, Lapland	77					174	
Bunting, Little	76						
Bunting, Ortolan	75					174	
Bunting, Pine	75						
Bunting, Pine X							
Yellowhammer	75						
Bunting, Red-headed	76						
Bunting, Reed	77	103	106		145	174	
Bunting, Rock	75						
Bunting, Rustic	76	103					
Bunting, Siberian Meadow	75						
Bunting, Snow	77	103				174	
Bunting, Yellow-breasted	76						
Bushchat, Rufous	58					169	
Bustard, Great	28	95		118	135	156	
Bustard, Houbara		95				157	
Bustard, Little	28	95				156	
Butcherbird, Grey			107				
Button-quail, Painted			107				
Buzzard	21	93			134	154	
Buzzard, Honey	19			114		153	
Buzzard, Long-legged	21						
Buzzard, Rough-legged	21	93		115		154	
Canary	78						
Canvasback	17						
Capercaillie	24	94	105	116		155	181
Cassowary, sp							178
Chaffinch	77		106	129	145	174	
Chaffinch, Blue	77						
Chickadee, Black-capped	73						
Chiffchaff	69				143	172	
Chough	87	103		131	146	176	
Chough, Alpine	87						
Chukar	25						
Cisticola, Zitting	71						
Coot	28			118	135	156	
Coot, Crested	28						
Cormorant	12			110	132	148	

	BS	MC	MM	MU	EM	EC	SK
Cormorant, Pygmy	13						
Corncrake	27			117	135	156	
Courser, Cream-coloured						157	
Crake, Baillon's	27	94				156	
Crake, Little	27					156	
Crake, Spotted	27			117		156	
Crane		94				155	
Crane, Demoiselle	26					155	
Crane, sp							181
Crossbill	82			130		175	
Crossbill, Parrot	81						
Crossbill, Scottish	82						
Crossbill, Two-barred	82						
Crow, Carrion	87	104		131	146	176	182
Crow, Carrion X Raven	87						
Crow, Collared	88						
Cuckoo	43			124	138	163	
Cuckoo, Fan-tailed			107				
Cuckoo, Great Spotted						163	
Cuckoo, Yellow-billed	43					164	
Cuckoo-shrike, Yellow-eyed			107				
Curassow, Great							181
Curlew	31	96			136	158	182
Dipper	57			126	141	168	
Diver, Black-throated	9			109		147	179
Diver, Great Northern	9			109		147	
Diver, Red-throated	9			108		147	179
Dodo							182
Dollarbird			107				
Dotterel	30	96				158	
Dove, Emerald			107				
Dove, Rock	41			122	138	163	
Dove, Rufous Turtle	42						
Dove, Spotted	43						
Dove, Stock	42			122	138	163	
Dove, Turtle	42			122	138		
Drongo, Greater Racket-tailed				130			
Duck, Ferruginous	17					152	
Duck, Indian Whistling	15						

	BS	MC	MM	MU	EM	EC	SK
Duck, Long-tailed	18	92				152	
Duck, Tufted	17			113		152	
Dunlin	34	97	105	120	137	159	
Dunnock	58		106	127	142	169	
Eagle, Bonelli's	22						
Eagle, Booted	22						
Eagle, Golden				115		154	
Eagle, Great Spotted	22					154	
Eagle, Short-toed	20						
Eagle, White-tailed Sea	20			114		153	180
Egret, Cattle						149	
Egret, Great White	14	91				149	
Egret, Little	13					149	
Eider	17		106	113	133	152	
Eider, King	17					152	
Eider, Spectacled	18						
Emu						177	
Falcon, Eleonora's	23						
Falcon, Red-footed	23	93				154	
Fieldfare	64					170	
Finch, Brandt's Mountain	80						
Finch, Citril	78						
Finch, Desert	80						
Finch, Gouldian			107				
Finch, Hodgson's Mountain	80						
Finch, Rosy	80						
Finch, Snow	85						
Finch, Trumpeter	80						
Firecrest	71					172	
Flamingo, Greater						150	
Flycatcher, Collared	71						
Flycatcher, Pied	71			128		172	
Flycatcher, Red-breasted	71					172	
Flycatcher, Spotted	72				144	172	
Flycatcher, Yellow-rumped	71						
Fulmar	10	90		109	132	147	
Gadwall						151	
Gallinule, Purple	27	95				156	

	BS	MC	MM	MU	EM	EC	SK
Game Cock, Old English		94					
Gannet	12			110		148	179
Garganey	16					151	
Godwit, Bar-tailed	30		105	119		158	
Godwit, Black-tailed	30			119		158	
Goldcrest	71		105	128		172	
Goldeneye	18	92				152	
Goldfinch	79	103	106	130		175	
Goldfinch X Canary	79						
Goosander	19	92	105	113	133	153	
Goose, Barnacle				111			
Goose, Bean						150	
Goose, Brent	15			111			
Goose, Canada		92					
Goose, Egyptian	15						
Goose, Greylag	15			111		150	180
Goose, Lesser White-fronted	15						
Goose, Magellan				111			
Goose, Pink-footed				111		150	
Goose, White-fronted				111		150	
Goshawk	21			114		153	
Grebe, Black-necked	10					147	
Grebe, Great Crested	10			109	132	147	
Grebe, Little	9	89	105		132	147	179
Grebe, Red-necked	9			109		147	
Grebe, Slavonian	10		105	109		147	
Greenfinch	78		106		145	174	
Greenfinch, Oriental	78						
Greenshank	31					159	
Grosbeak, Pine	81			130		175	
Grouse, Willow / Red	24	93		116	134	155	
Grouse, Black	24			116		155	
Grouse, Black X Capercaillie			105				
Grouse, Hazel	25						
Guan, sp							181
Guillemot	40	99		122	138	162	
Guillemot, Black	41	99		122		162	
Guillemot, Brunnich's						162	
Gull, Black-headed	37	98		121	137	161	
Gull, Bonaparte's	38					161	

	BS	MC	MM	MU	EM	EC	SK
Gull, Common	36		106	120	137	160	
Gull, Franklin's	37						
Gull, Glaucous	37			121		160	
Gull, Great Black-backed	37			121		160	182
Gull, Great Black-headed						160	
Gull, Herring	36	98	106		137	160	
Gull, Iceland	37			121		160	
Gull, Ivory	36						
Gull, Lesser Black-backed	36			120	137	160	
Gull, Little	38	98				161	
Gull, Mediterranean	37						
Gull, Pacific	36						
Gull, Ross's	38						
Gull, Sabine's	38	98					
Gull, Slender-billed	38						
Gull, Yellow-legged	36						
Gyrfalcon	23			115		154	
Harlequin	18					152	
Harrier, Hen	20			114		153	
Harrier, Marsh	20			114	133	153	181
Harrier, Montagu's	21				134	153	
Harrier, Pallid	20						
Hawfinch	83			130	145	175	
Hemipode, Andalusian	26						
Heron, Green						149	
Heron, Grey	14			110	133	149	179
Heron, Night	13	91				149	
Heron, Purple	14			110		149	179
Heron, Squacco	13	91				149	
Hobby	23	93		115	134	154	
Hoopoe	47			125	140	166	
Hornbill, Great Indian				125			
Hornbill, Rhinoceros				125			
Hornbill, Wreathed				125			
Ibis, Glossy	14					150	
Jackdaw	87	103		131	146	176	
Jay	86	103		130	146	176	

	BS	MC	MM	MU	EM	EC	SK
Kakapo			106				
Kestrel	23	93		115	134	154	
Kestrel, American	22						
Kestrel, Lesser				115		154	
Killdeer						158	
Kingfisher	46	101			140	165	
Kingfisher, Black-headed	46						
Kingfisher, Forest			107				
Kinglet, Golden-crowned	70						
Kinglet, Ruby-crowned	70						
Kite, Black	19					153	
Kite, Black-shouldered	19						
Kite, Red	20	93		114		153	
Kittiwake	38		106	121	137	161	
Kiwi, Brown			106	108			
Kiwi, Great Spotted				108			179
Kiwi, Little Spotted / Great Spotted			106	108			
Knot	33		105	119			
Koel			107				
Lapwing	29			118	135	157	
Lark, Black	50						
Lark, Calandra	49						
Lark, Crested	50					166	
Lark, Lesser Short-toed	50						
Lark, Shore	51					167	
Lark, Short-toed	50					166	
Lark, White-winged	49					166	
Linnet	80				145	175	
Lory, Blue-streaked				123			
Lory, Papuan				123			
Lory, Rainbow				123			
Magpie	86	103		131	146	176	
Magpie, Azure-winged	86						
Mallard	16			112	133	151	180
Mallard X Pintail				112			
Martin, Crag	51						
Martin, House	51	101			140	167	
Martin, Sand	51	101	106		140	167	

	BS	MC	MM	MU	EM	EC	SK
Partridge, Grey	25		106		134		
Partridge, Red-legged	25			116	134	155	
Peafowl				117			181
Pelican, sp							179
Penguin, Jackass							179
Peregrine	24		105	116		155	
Petrel, Bulwer's		90		109		148	
Petrel, Leach's	12	90				148	
Petrel, Storm	12	90				148	
Petrel, White-faced Storm						148	
Petrel, Wilson's	11						
Phalarope, Grey	32	96				159	
Phalarope, Red-necked	32	96				159	
Pheasant	26	94	105	117			
Pheasant, Golden	26			117			
Pheasant, Reeve's		94	105				
Pigeon, Blue-crowned				122			
Pigeon, Passenger	42	99					
Pintail	16			112		151	
Pipit, Berthelot's	54						
Pipit, Buff-bellied	54						
Pipit, Hodgson's	53						
Pipit, Meadow	53	102			141	167	
Pipit, Olive-backed	53						
Pipit, Red-throated	53					168	
Pipit, Richard's	53						
Pipit, Rock	54				141	168	
Pipit, Tawny	53					167	
Pipit, Tree	53				141	168	
Pipit, Water	54						
Plover, Caspian	30						
Plover, Egyptian	29						
Plover, Golden	29	96			135	158	
Plover, Grey	29	96					
Plover, Kentish	30		105			158	
Plover, Little Ringed	30					158	
Plover, Pacific Golden	29						
Plover, Ringed	30		105		136	158	
Plover, Sociable						158	
Pochard	17			112		152	
Pochard, Red-crested	17			112		151	

	BS	MC	MM	MU	EM	EC	SK
Pratincole, Collared	29			118		157	
Ptarmigan	24		105	116		155	
Puffin	41	99			138	162	
Quail	25					155	
Quail, Japanese	25						
Rail, Water	26	94		117		155	
Raven	88			131		176	
Razorbill	40	99		121	138	162	
Redpoll	79			130	145	175	
Redpoll, Arctic	79						
Redshank	31	96			136	159	
Redshank, Spotted	31			119			
Redstart	60			127	142	169	
Redstart, Black	59	102					
Redstart, Daurian	60						
Redstart, Moussier's	60						
Redstart, Plumbeous	60						
Redwing	64			128		170	
Riflebird, Queen Victoria	86						
Robin	58	102	106	127	142	169	
Robin, Eastern Yellow			107				
Robin, Rose			107				
Robin, White-throated	59						
Roller	47					165	
Rook	87	104		131	146	176	
Rosefinch, Long-tailed	81						
Rosefinch, Pallas's	81						
Rosefinch, Scarlet	81						
Rosefinch, Vinaceous	81						
Rubythroat, Siberian	59						
Ruff	35	97	105	120	137	160	
Saker				116			
Sanderling	34	97	105				
Sandgrouse, Black-bellied	41						
Sandgrouse, Pallas's	41	99				163	
Sandgrouse, Pin-tailed	41						
Sandpiper, Baird's	34						
Sandpiper, Broad-billed						160	

	BS	MC	MM	MU	EM	EC	SK
Sandpiper, Common	32			119	136	159	
Sandpiper, Curlew	35	97					
Sandpiper, Green	31	96		119			
Sandpiper, Marsh	31						
Sandpiper, Pectoral	34	97					
Sandpiper, Purple	34			119		159	
Sandpiper, Spoon-billed	35						
Sandpiper, Spotted	32						
Sandpiper, Terek	32						
Sandpiper, Upland						158	
Sandpiper, Wood	31	96				159	
Scaup	17	92		113		152	
Scoter, Common	18			113		152	
Scoter, Velvet	18			113		152	
Serin	78					174	
Serin, Red-fronted	78						
Shag	12	91		110	133	149	
Shearwater, Balearic		90					
Shearwater, Great	10					148	
Shearwater, Little	11					148	
Shearwater, Manx	11	90		110		148	
Shearwater, Sooty	10			110		148	
Shelduck	15			112		151	
Shelduck, Ruddy						151	
Shoveler	16			112		151	
Shrike, Black-headed Bush	54						
Shrike, Brown	55						
Shrike, Bull-headed	55						
Shrike, Chinese Great Grey	56						
Shrike, Great Grey	56					168	
Shrike, Isabelline	55						
Shrike, Lesser Grey	56					168	
Shrike, Loggerhead	56						
Shrike, Masked	57						
Shrike, Red-backed	55				141	168	
Shrike, Rufous-backed or Long-tailed	55						
Shrike, Tiger	55						
Siskin	79			129		174	
Skua, Arctic	35		106	120		160	
Skua, Great	35					160	

	BS	MC	MM	MU	EM	EC	SK
Skua, Long-tailed	36			120		160	
Skua, Pomarine	35			120			
Skylark	50	101			140	166	
Smew				113			
Snipe	33	97			136	159	
Snipe, Great	33	97					
Snipe, Jack	33	97				159	
Snipe, Pin-tailed	33						
Sora	27						
Sparrow, Cinnamon	84						
Sparrow, Dead Sea	84						
Sparrow, House	83		106	130	145	175	
Sparrow, House / Spanish	84						
Sparrow, Rock	84						
Sparrow, Saxaul	84						
Sparrow, Spanish	83						
Sparrow, Tree	84		106	130	146	175	
Sparrowhawk	21			114	134	154	
Spoonbill	14	92				150	180
Spoonbill, Yellow-billed	15						
Starling	85		106	130	146	175	
Starling, Rose-coloured	85					175	
Starling, Spotless	85						
Stilt, Black-winged						157	
Stint, Little	34						
Stint, Temminck's	34	97			136	159	
Stone-curlew	29	95		118	135	157	
Stonechat	60			127	142	169	
Stonechat, Pied	61						
Stonechat, White-tailed	60						
Stork, Black		91				150	
Stork, Marabou							180
Stork, White						150	180
Swallow	51	101	106		140	167	
Swallow, Red-rumped	51						
Swan, Bewick's				111			180
Swan, Mute	15			111		150	
Swan, Whooper				111		150	180
Swan, sp							180
Swift	45	100			139	165	
Swift, Alpine	45	100					

212

	BS	MC	MM	MU	EM	EC	SK
Swift, Little or House	46						
Swift, Pallid	45						
Swift, Plain	45						
Teal	16			112	133	151	
Teal, Baikal	16						
Teal, Blue-winged						151	
Teal, Marbled	16						
Tern, Arctic	39				137	161	
Tern, Black	38	98				161	
Tern, Caspian	39					161	
Tern Common	39		106		137	161	
Tern, Common / Arctic					137		
Tern, Gull-billed	39					161	
Tern, Lesser Crested	40						
Tern, Little	39	98			137	162	
Tern, Roseate					137	161	
Tern, Royal	40						
Tern, Sandwich	40				138	162	
Tern, Sooty						162	
Tern, Whiskered						161	
Tern, White-winged Black	38					161	
Thrush, Australian Ground	63						
Thrush, Black-throated	64						
Thrush, Blue Rock	63						
Thrush, Dusky	64						
Thrush, Eye-browed	64						
Thrush, Mistle	64				143	170	
Thrush, Rock	63					170	
Thrush, Song	64			128	142	170	
Thrush, White's		102		127			
Tit, Bearded						172	
Tit, Blue	73		106	129	144	173	
Tlt, Coal	73				144	173	
Tit, Crested	73					173	
Tit, Great	73			129	144	173	
Tit, Long-tailed	72			129	144	172	
Tit, Marsh	72			128	144	173	
Tit, Penduline	72						
Tit, Sombre	72						
Tit, Willow	73				144		

	BS	MC	MM	MU	EM	EC	SK
Toucan, Toco				126			
Treecreeper	74			129	144	173	
Treecreeper, White-throated			107				
Turkey							181
Turnstone	32		105	119	136	159	
Twite	80					175	
Vulture, Egyptian						153	
Vulture, Griffon						153	181
Vulture, King							180
Vulture, Palm-nut							181
Wagtail, Citrine	52						
Wagtail, Grey	52	102			141	167	
Wagtail, Large Pied	52						
Wagtail, Madagascar	52						
Wagtail, Pied	52				141	167	
Wagtail, Yellow	52				141	167	
Wallcreeper	74						
Warbler, Aquatic	65					170	
Warbler, Arctic	70						
Warbler, Barred	67					171	
Warbler, Black-browed Flycatcher	70						
Warbler, Black-browed Reed	66						
Warbler, Blyth's Reed	66						
Warbler, Cetti's	65						
Warbler, Cyprus	68						
Warbler, Dartford	69				143	171	
Warbler, Desert	68						
Warbler, Dusky	69						
Warbler, Fan-tailed	71						
Warbler, Garden	67			128	143	171	
Warbler, Grasshopper	65				143	170	
Warbler, Great Reed	66					171	
Warbler, Grey-headed Flycatcher	70						
Warbler, Icterine	66					171	
Warbler, Lanceolated	65						
Warbler, Marsh	66					171	

	BS	MC	MM	MU	EM	EC	SK
Woodcock	33			119	136	159	
Woodlark	50	101			140	166	
Woodpecker, Great Spotted	48			126		166	
Woodpecker, Green	49	101		126		166	
Woodpecker, Grey-headed	49						
Woodpecker, Japanese Green	49						
Woodpecker, Lesser Spotted	47			126			
Woodpecker, Middle Spotted	48						
Woodpecker, Pileated	48						
Woodpecker, Syrian	48						
Woodpecker, Three-toed	48						
Woodpecker, White-backed	47						
Woodpigeon	42			122	138	163	
Wren	58			127	141	168	
Wryneck	47			126	140	166	
Yellowhammer	74		106		145	173	
Yellowlegs, Lesser		96